T0413940

Schooling for Sustainable Development in South America

Schooling for Sustainable Development

Volume 2

This book series addresses issues associated with sustainability with a strong focus on the need for educational policy and action. Current attention and initiatives assume that Education for Sustainable Development (ESD) can be introduced successfully and gradually into schools worldwide. This series explores the issues that arise from the substantial and sustainable changes to be implemented in schools and education systems.

The series aims to counter the prevailing Western character of current research and enable cross-cultural comparisons of educational policy, practice, and project development. As a whole, it provides authoritative and comprehensive global coverage, with each volume providing regional/continental coverage. The volumes present data and insights that contribute to research, policy and practice in ESD-related curriculum development, school organization and school-community partnerships. They are based on ESD-related project experiences, empirical studies that focus on ESD implementation and teachers' perceptions as well as childhood studies that examine children's geographies, cultural characteristics and behaviours.

For further volumes:
http://www.springer.com/series/8635

Maria Lucia de Amorim Soares
Leandro Petarnella
Editors

Schooling for Sustainable Development in South America

Policies, Actions and Educational Experiences

Springer

Editors
Maria Lucia de Amorim Soares
University de Sorocaba - UNISO/SP
Avenida Dr. Eugenio Salerno
100/140
18035-430 Santa Terezinha -
Sorocaba-SP
Brazil
maria.soares@prof.uniso.br

Leandro Petarnella
University Nove de
Julho-Uninove/SP
Rua Tupi 33, apto 2
01233-001 Santa Cecilia-
Sao Paulo-SP
Brazil
leandro_66@hotmail.com

ISBN 978-94-007-1753-4 e-ISBN 978-94-007-1754-1
DOI 10.1007/978-94-007-1754-1
Springer Dordrecht Heidelberg London New York

Library of Congress Control Number: 2011935370

Printed on acid-free paper

Springer is part of Springer Science+Business Media (www.springer.com)

For Michael Williams, John Chi-Kin Lee and Philip Stimpson who, trusting us, invited us to organise this book without imposing anything or compelling us to think outside the pleasure of thought.

For those who we invited to write their chapters and who articulated a set of arguments that turned inside out education for sustainable development in South America.

For our readers who, when they respond to the question "What can I do?", may have as their goal the importance of life, freedom and creativity in the world.

Maria Lucia de Amorim Soares
and Leandro Petarnella

Acknowledgements

Writing and co-editing a book is a risky task. You do this and present the ideas, concepts and inconveniences that involve a certain area giving concreteness to what is transitory in daily life. For this to happen, in this book many people were involved and we would like to thank everyone who took this risk with us.

Special thanks to Emeritus Professor Michael Williams, formerly of the Faculty of Education and Health Studies from Swansea University, in the United Kingdom, who besides believing in our work, worked hard in reviewing all of the chapters in this book. Also thanks to Professor John Chi-Kin Lee of the Hong Kong Institute of Education, who, with Dr Philip Stimpson, formerly of the University of Hong Kong and Professor Michael Williams, together have offered this challenge to us: the mission of bringing together the research findings, experiences and political debates taking place in universities in South America about the environment, education and sustainable development.

We also thank all the authors who together helped make this work: Vânia Regina Boschetti, Fabián Araya Palacios, Ubiratan Silva Alves, Maria Beatriz Rocha Ferreira, Cláudia Marcela Polimeni, Rosa López de D'Amico, Maritza Loreto and Orlando Mendoza who, contributing the chapters in the first part of this book, helped us draw a general overview of educational programmes and policies that have been adopted in South America, showing that sustainable development can become a reality in the countries they represent.

Our thanks go also to Sergio Luiz de Souza Vieira, Carmen Lucia Artioli Rolim, José Damião Trindade Rocha, Paulo Alexandre Adler Pereira and Eduardo de Campos Garcia, who, in the second part of this work, enabled us to understand the work that is being developed in Brazil involving sustainability.

Finally, we are grateful to Nara Silvia Marcondes Martins, Ivo Eduardo Roman Pons, Petra Sanchez Sanchez, Eliete Jussara Nogueira, Luiz Fernando Gomes da Silva and Paulo Celso da Silva who, in the third part, have called on their educational experiences, to reveal how an education for sustainable development can go beyond the curricular discourse.

Series Editors' Introduction

Education for sustainable development (ESD) has rapidly become part of educational discourses worldwide. Within its global attractiveness lie both its strength and its weakness. Its strength lies in its capacity to alert educationists, broadly defined, to a shared concern for the future of both the planet and local communities. Its weakness lies in its lack of shared meaning and, stemming from this, the enormous difficulties encountered in trying to bring ESD into the mainstream activities of educational institutions.

In designating the period 2005–2014 as the International Decade of Education for Sustainable Development the United Nations sought to bring to the fore the need for politicians, policy-makers and practitioners to seek ways by which ESD can become part of the fabric of formal and informal education. At the heart of the numerous initiatives that have been stimulated by this designation is the assumption that ESD should be introduced and can be introduced successfully into schools world-wide. It is assumed that children, older students and adults can be educated formally to act now in the interests of a sustainable future and to act internationally.

What is evident is that different nations have adopted different approaches to ESD, sometimes interchanging the term with environmental education, another term subject to a wide range of interpretations. These differences are evident in educational practice in regions, districts and individual schools as well as in academic studies and commentaries. Obviously, this is not to say that there is some common ground in policies and practice, it is simply to keep to the forefront the recognition that, even when nations make pronouncements about aspects of ESD, these should not be treated as authoritative statements about what is happening at the school and classroom levels. Broad statements have a value in highlighting issues and trends but they need to be treated with caution. The same caution needs to be applied to pronouncements emanating from academic sources. Academics have their own agendas and care must be taken when reading what appear to be authoritative statements about developments in ESD occurring within their own communities and nations.

Our series addresses the array of issues arising from attempts made to convert assumptions about, and definitions of, ESD into substantial and sustainable changes principally in schools. Underpinning the series is a concern for identifying those cultural forces that impact on national, regional and local adaptations to approaches to ESD that have international currency. In this, the editors of the books in the series, each based on experience in a single continent or extensive region, seek to counter the strong Western (Australian, North American, European) character of much research and writing in the broad field of ESD. Research and scholarly studies are commonly underpinned by values and assumptions derived from Western culture, broadly defined. The design of the series as a set of broadly continent-scale books seeks to bring together experts from various countries in each continent. The books bring out contrasting experiences and insights with a range of explanations of policies and practice.

Within the broad cultural contexts of the continents and regions included in the series authors provide evidence of policies, formal curriculum developments and innovations and informal school-related activities. Some authors have paid close attention to policy making at various levels, others have addressed whole school organisational issues and others have provided detailed case studies of localities and individual schools.

Children and young people live in distinct worlds of their own. They have very distinctive cognitive and affective characteristics that vary from one culture to another, at whatever scale that culture is defined. They are also often targets for environmental campaigns that wish to promote particular behavioural changes. ESD is often construed as an attempt to change habits, to encourage children and young people to "think globally and act locally". This series demonstrates how this and other slogans are translated in education systems and schools world-wide.

For this volume *Schooling for Sustainable Development in South America: Policies, Actions and Educational Experiences*, the co-editors Maria Lucia de Amorim Soares and Leandro Petarnella have brought together an array of chapters highlighting the recent developments and issues related to ESD in selected South American nations. The book is divided into three parts. Part I gives an overview on schooling for sustainable development in South America. Part II focuses on schooling for sustainable development in Brazil while Part III depicts the trends and challenges in educational provision for sustainable development. ESD is emphasised with a socio-political orientation in the context of developing countries in South America and in particular in Brazil where many scholars locally and internationally draw upon the inspiring ideas of Paulo Freire who called for social and educational actions in eradicating social injustice. In some countries like Bolivia and Brazil, as described in Chaps. 6 and 8 respectively, while environmental education or ESD is still an emerging and important field, perhaps the fundamental challenge of ESD remains whether or not to make an overhaul of the whole education system the top priority as an attempt to reduce illiteracy rates and help students ensure their own survival. There are important lessons for the governments in South America for future improvement and calls for international attention and co-operation.

John Chi-Kin Lee
Michael Williams
Philip Stimpson

Contents

Part I Schooling for Sustainable Development in South America

1 **Schooling for Sustainable Development: Autonomy, Citizenship and Social Justice in South America** 3
Maria Lucia de Amorim Soares and Leandro Petarnella

2 **Sustainability and Educational Actions** 19
Vânia Regina Boschetti

3 **Schooling for Sustainable Development in Chile** 33
Fabián Araya Palacios

4 **Education and Sustainability: Figurations and the Ethics of Care – Bolivian Experience** 53
Ubiratan Silva Alves and Maria Beatriz Rocha Ferreira

5 **Geography and Sustainable Development in Teaching and Education in Argentina** 71
Cláudia Marcela Polimeni

6 **Venezuela and Education Transformation for the Development of the People** 87
Rosa López de D'Amico, Maritza Loreto, and Orlando Mendoza

Part II Schooling for Sustainable Development in Brazil

7 **The Environment as a Cross-Curricular Theme in the Brazilian "Parâmetros Curriculares Nacionais" (PCNs, National Curriculum Parameters)** 105
Maria Lucia de Amorim Soares and Leandro Petarnella

8 **School Culture in Brazil: Complexities in the Construction of the Sustainability Concept** 123
Sergio Luiz de Souza Vieira

9 Schooling for Sustainable Development: Experience with Students of Early Childhood Education in the State School Paulo Tapajós, São Paulo/Brazil 143
Maria Lucia de Amorim Soares, Leandro Petarnella, and Eduardo de Campos Garcia

10 Higher Education and Regional Sustainable Development: The Case of the Federal University of Tocantins in the Brazilian Cerrado 159
Carmen Lucia Artioli Rolim, José Damião Trindade Rocha, and Paulo Alexandre Adler Pereira

Part III Trends and Challenges of Educational Provision for Sustainable Development

11 Social and Environmental Design as an Educational Pedagogic Resource for Sustainable Development: An Experience with NGOs and Universities 181
Nara Silvia Marcondes Martins, Ivo Eduardo Roman Pons, and Petra Sanchez Sanchez

12 Education for Sustainability, Ethical Relations 193
Eliete Jussara Nogueira

13 Digital Literacy and Sustainability: The *Vozes que Ecoam* Project 205
Luiz Fernando Gomes

14 At School You Learn About the City: Urbanisation and Sustainable Development in Education 219
Paulo Celso da Silva

Index 231

Contributors

Ubiratan Silva Alves His undergraduate studies were in Physical Education at the University of São Paulo followed by a Masters degree in the field of Psychology and Education from the Faculty of Education, University of São Paulo. He is currently studying for a Ph.D. in the Faculty of Physical Education, State University of Campinas (UNICAMP/SP). He has been coach and trainer of the São Paulo Futebol Clube. He has taught higher education teaching courses in Physical Education and Education since 1997. His research and various publications are in the areas of Physical Education, the Physical Education professional, undergraduate sports, football, anthropology, recreation, leisure and culture.

Rosa López de D'Amico She has a Masters in Education and a Ph.D. in Sports Management. She is a member of the Research Promotion Programme (IPP – Level III). She has served as a Coordinator of the Research Centre for Studies in Physical Education, Health, Sports, Recreation and Dance and Coordinator of the Socio-Cultural Extension Programme. She is Head of Literature and is Assistant Director of Research. She has received national and international awards, most recently in Academic Productivity awarded by the Nucleus of Scientific, Humanistic and Technological Universities Venezuela (CDCHT).

Vânia Regina Boschetti She is a graduate in Philosophy from the University of Sorocaba (1971) and she also has a degree in Social Studies from the University of Tatuí (1977). She graduated in pedagogy from the University of Sorocaba in 1983 and holds a Masters degree in Education from the Universidade Metodista de Piracicaba (1992) and a Ph.D. from the University of São Paulo. She is currently a Professor in the Doctoral Education Programme at the University of Sorocaba (Uniso) where she has special interests in education, learning, schools, basic schooling and reading.

Maria Beatriz Rocha Ferreira She is a Professor in the Faculty of Physical Education at the University of Campinas (UNICAMP/SP) with a Ph.D. in Anthropology from the University of Texas, Austin (1987) and a Masters degree from the State University of São Paulo (USP/SP) (1972 and 1980). She is a member of the Faculty of Physical Education at State University of Campinas (UNICAMP/SP) where she is Head of the

Laboratory for Bio-cultural Anthropology developing projects on physical activity, games and sports. She is a Professor in the Graduate School of Physical Education, State University of Campinas (UNICAMP/SP).

Eduardo de Campos Garcia He has a Masters in Education, Art and Culture from Mackenzie University, Brazil. He is engaged in Specialist Teaching in Higher Education in the Catholic University of São Paulo (PUC-SP) with special interests focused on sign language and the teaching and learning of the deaf in literature.

Luiz Fernando Gomes He has an undergraduate degree from the University of Sorocaba (1991), a Masters degree in Linguistics from the Catholic University of São Paulo (PUC-SP) (1998) and a Ph.D. in Linguistics from University Estadual de Campinas (2007). His specialist interest in linguistics has been applied to studies of distance education and new technologies in education.

Maritza Loreto He has a Doctorate and a Master's degree in Education. He is currently a Professor at the Pedagogical University Experimental Libertador – Maracay Educational in Venezuela, having worked in the Ministry of Education of the State of Aragua and as Coordinator of Science in Mission Aragua. He has experience in education with an emphasis on educational policies.

Nara Silvia Marcondes Martins She is a graduate in Fine Arts from the Faculty of Fine Arts of São Paulo (1990). She has an MA in Visual Arts at the Arts Institute of the State University of São Paulo (UNESP) (1996) and a Ph.D. in Architecture and Urbanism, Faculty of Architecture and Urbanism, University of São Paulo (USP-FAU) (2001). She is currently a Professor and Researcher at Mackenzie University and Coordinator of the Industrial Design course at the Faculty of Architecture and Urbanism, Universidade Presbiteriana Mackenzie. She has experience in socio-environmental design.

Orlando Mendoza He is a Graduate in Mathematics and has a Masters in Education. He is currently Consultant to the National Council for Studies Science and a Professor at the Pedagogical University Experimental Libertador (UPEL) in Maracay, Venezuela. He is a member of the core research group in mathematics education from the Centre of Pedagogical University Experimental Libertador – Maracay.

Eliete Jussara Nogueira She graduated in Psychology at the Catholic University of Campinas (1985) and has a MA in Education from the State University of Campinas (1992) and a Doctorate in Education from the State University of Campinas (2001). She has experience in education with an emphasis on Educational Psychology and Human Development with special interests in social networks, aging, relationships, teaching practices, the contemporary world and everyday school life.

Fabián Araya Palacios He has a Teacher Qualification in History and Geography and his Bachelor of Education from the University of La Serena, Chile (1990). He obtained his Master of Education in Geography from the University Pedagógica Nacional de Colombia, Colombia, (1996) and his Ph.D. in Geography from the University Nacional de Cuyo, Mendoza, Argentina (2006). Professor Palacios is the co-director of national and international research projects and is a geographical education advisor for the Ministry of Education of Chile. His research interests focus on Geographic Education for Sustainability of the Rural Environment.

Paulo Alexandre Adler Pereira He graduated in Philosophy at the University of Rio de Janeiro State (UERJ), Rio de Janeiro, Brazil and in Physics at the University of Rio de Janeiro State (UERJ), Rio de Janeiro, Brazil. He has a Masters in Philosophy of Science from the Federal University of Rio de Janeiro (UFRJ) Rio de Janeiro, Brazil. He was awarded a Ph.D. in the theory of knowledge by the Pontiff Catholic University of Rio de Janeiro (PUC/RJ), Rio de Janeiro, Brazil. He is an Assistant Professor for the Pedagogy course at the Federal University of Tocantins (UFT), Palmas, Brazil.

Leandro Petarnella He has a Doctorate in Education from the University of Sorocaba and Doctorate in Administration from University Nove de Julho. He was awarded a Master of Education by the University of Sorocaba (2008) and graduated in Mathematics from the University Bandeirante of São Paulo (2005) and Accounting from University Nossa Senhora do Patrocínio (2002). He is currently a Professor at the University Nove de Julho and has researched aspects of schools with regard to distance education, technologies applied to education and sustainability education in undergraduate courses.

Cláudia Marcela Polimeni She graduated in Geography at the University Nacional of Cuyo, Argentina, where she also gained her Doctorate in Geography. She is currently a Professor at the University Nacional of Cuyo with research interests focused on Geographical Education for Sustainability of Human Resources.

Ivo Eduardo Roman Pons He is a graduate in Industrial Design with specialisation in Product Design from Mackenzie University (2001) and has a Masters in Education, Art and Cultural History and a Ph.D. in Architecture and Urbanism from Mackenzie University (2006). He is currently President of the NGO Design Possible, an Assistant Professor at Mackenzie University and volunteer adviser at the Centre for Design and Fashion NGO. He has experience in Industrial Design, with an emphasis on product design and product data management and production. He is a researcher in the area of social and sustainable design.

José Damião Trindade Rocha He graduated in Pedagogy at the University of Amazon (UNAMA), Belém. He is a Specialist in Higher Education Teaching with special interests in Education, Communication and New Technology. He has a Master in Education from the Federal University of Goiás (UFG) and a Ph.D. in Education from the Federal University of Bahia (UFBA), Salvador. He is an Assistant Professor for the Pedagogy course at the Federal University of Tocantins (UFT).

Carmen Lucia Artioli Rolim She graduated in Science and Mathematics and she is a specialist in Computer Science and Information Systems. She gained her Master in Education from UNISO (University of Sorocaba) and her Ph.D. from UNIMEP (Methodist University of Piracicaba). She is an Assistant Professor for the Pedagogy Course at UFT (Federal University of Tocantins). She has experience in education with an emphasis on methods and techniques of teaching.

Petra Sanchez Sanchez He is a postgraduate in Public Health from the University of São Paulo (USP) and has a Doctorate in Science/Microbiology from the Institute of Biomedical Sciences University of São Paulo (USP). He is currently a professor in the Graduate Programme in Education, Art and Cultural History and a researcher at Mackenzie University. He has experience in the area of sustainability as a scientific advisor in government institutes in Brazil and Latin America, such as the Pan American Health Organization – CEPIS, Research Support Foundation of São Paulo (FAPESP).

Paulo Celso da Silva He graduated in Geography (1988) and gained a Ph.D. from the Faculty of Philosophy and Letters of Sorocaba (1989) and gained a Ph.D. He holds a Masters degree in Geography (Human Geography) from the University of São Paulo (1995) and was awarded a Ph.D. in Geography (Human Geography) by the University of São Paulo (2000). He engaged in post-doctoral studies at the University of Barcelona (2001–2002). He is currently a Professor at the University of Sorocaba. His specialist interests are in Communication Geography and Philosophy with an emphasis on communication in the contexts of learning, communication and the city, media and urban social movements.

Maria Lucia de Amorim Soares She graduated from the Faculty of Philosophy of Sciences and Letters Itapetininga (1973) with a Bachelor of History and Geography from the Faculty of Philosophy Sciences and Letters Sapiential (1958). She has a Masters degree in Geography (Human Geography) from the University of São Paulo (1981) and a Ph.D. in Science (Human Geography) from the University of São Paulo (1996). She is currently a Professor at the University of Sorocaba, a collaborator in the Ford Foundation Carlos Chagas, Brazilian Society Science Progress – SP and the State University of Campinas. She has experience in education, with an emphasis on methods and techniques of teaching, with particular reference to environmental education, geography and postmodernism.

Sergio Luiz de Souza Vieira His undergraduate studies were in Physical Education at the University of Guarulhos (1989) and at the same institution he followed specialist courses in Physical Education and Children's Gymnastics (1990). He holds a Ph.D. and Master of Social Sciences (Anthropology) from the Catholic University of São Paulo (PUC/SP) (2004 and 1997) with studies in Afro-Brazilian Culture with reference to religion, memory and identity. He is engaged in research on society, culture and environment, analysing the use of wood taken from native forests to meet the demands of tourism, folklore and sports and the contradictions between public policies of environmental preservation and protection of cultural heritage.

Part I
Schooling for Sustainable Development in South America

Chapter 1
Schooling for Sustainable Development: Autonomy, Citizenship and Social Justice in South America

Maria Lucia de Amorim Soares and Leandro Petarnella

Introduction

South America is a continent that exhibits a great diversity in natural and cultural phenomena. Much international environmental debate has focused on the challenges to the biodiversity and traditional communities of the Amazonian rainforests and the unique flora and fauna of the Galapagos Islands. However, each nation has its own particular environmental features that attract tourists and scientists from many countries. Historically, there are sharp contrasts between the experiences of indigenous communities and the immigrants who brought foreign values and practices that transformed economies and cultures. There has been an evolution from European colonial regimes through, in many South American nations, a series of mainly autocratic and often military dictatorships to the broadly democratic governments of today. As governments have changed so have educational policies and practices. Attempts to introduce and develop environmental education for sustainable development have to be viewed nation by nation in evolving political and economic contexts.

Environmental education is a process of political education which enables the acquisition of knowledge and skills as well as the formation of attitudes that that will become necessary for political citizenship and social justice with a view to ensuring a sustainable society. We should not lose sight of the complex ecological, social, economic and cultural challenges it presents. Education can add capabilities that can contribute in positive ways to the process of sustainable development,

M.L. de Amorim Soares (✉)
University of Sorocaba (UNISO/SP), Sorocaba, São Paulo, Brazil
e-mail: maria.soares@prof.uniso.br

L. Petarnella
University Nove de Julho, São Paulo, SP, Brazil

M.L. de Amorim Soares and L. Petarnella (eds.), *Schooling for Sustainable Development in South America*, Schooling for Sustainable Development 2,
DOI 10.1007/978-94-007-1754-1_1,

offering multiple alternatives for each locality, region or nation having cultural, environmental and ethical, moral values appropriate to their different stages of development.

Environmental education trains and prepares citizens for critical reflection and social action in order to make possible the full development of human beings. It occupies a position contrary to the prevailing model of economic development in which ethical values, social justice and solidarity are neglected. Education for sustainability offers an alternative position to profit at any price, competition, selfishness and the privileges of the few over the majority of the population.

The development of an environmental process involves a situational analysis. From this, educational goals to be achieved should be established, drawing on contributions from such different areas as geography, history, psychology, sociology, physics, economics and biology with their associated pedagogies that give rise to their school activities. The need for an interdisciplinary approach to the knowledge and understanding of environmental issues is essential for changing the images of environmental destruction throughout the world.

Education, as an activity present in all human societies, has as its basis the transmission of a society's cultural heritage to young people. Educational discourse incorporates the notion of transgenerational transmission, whether of knowledge, practices or elements traditionally considered sacred. Concomitantly, education also has the function to contribute to the transformation of systems that may be irrelevant or inadequate. Thus, education is seen as the biggest project for the survival of any society. Environmental education depends on systematic and co-ordinated actions that converge on the ethical commitment of mankind to life.

Building an agenda for the twenty-first century, an agenda for humanity, is an innovation that has not been thought of in other times. This is the project of environmental education, making the importance of environmental education something undeniable, assigning it an uncontested position in citizenship education and the promotion of social justice. It is worth noting, however, that environmental education has a recent history and that it is still maturing. But it should be clear that environmental education constitutes one of the ways of combating the environmental crisis, a strategy to be used against it. It is necessary, therefore, to think of environmental education as an instrument of citizenship formation and as a vehicle for creating a new alliance between mankind and nature.

Society needs to wake up to a twenty-first century world in which building a culture of sustainability is one of the main political and cultural challenges. What does it mean to be in the Decade of Education for Sustainable Development 2005–2014, as proposed by the United Nations? Carvalho argues that environmental education should pursue the path of active learning and not be identified only with changes in behaviour. This is because:

> A person can learn to value a healthy and unpolluted environment, to have behaviours such as not littering the streets and participating in neighbourhood clean-ups working with other people. That same person, however, may consider the production of an appropriate policy for the transfer of toxic waste to the region and not care about the contamination of a place far from their living environment (2001: 49).

Environmental education plays an important role in preparing this person. It is through environmental education that one encounters the possibilities of forming citizens who are aware of their autonomy and their citizenship. In the case of South America, it is through an education for sustainable development that it becomes possible to create new opportunities, including opportunities for social justice.

Brazil is one of the largest nations in the world, in both geographical area and population size. It is a nation that displays a vast gap between the opulence of the centres of such cities as Brasilia, Sao Paulo and Rio de Janeiro and the extensive slums which surround them. It also exhibits sharp contrasts between the globalised cultures of the wealthier citizens and the localized material poverty of indigenous peoples. The need for communities to address the importance of natural and physical environments in the context of sustainablity is evident in recent attempts to include environmental education and education for sustainable development in both national policies and local practices. In this introductory chapter, we have focused on Brazil since it highlights a number of educational features that are replicated elsewhere in South America.

Education for Sustainable Development: Pathways

In Brazil, in August 1981, Federal Law 6938 was sanctioned and provided for a National Environmental Policy that includes aims and the mechanisms for their formulation and execution. Environmental education is regarded as one of its foundations to be applied to all levels of education, including community education, to enable citizens to participate actively in environmental protection (Brazil 1981).

With regard to the inclusion of sustainable development, the Brazilian Constitution reflects contemporary thinking, as can be seen in Article 225,

> Everyone is entitled to an ecologically balanced environment and the common use and essential quality of healthy life, imposing upon the State and society the duty to defend and preserve it for present and future generations (Brazil 1988).

According to Federal Law 9795, 1999, which provides for an Environmental Education Policy, everyone has the right to environmental education as an essential and permanent component of national education. It is to be exercised in a coordinated way at all levels and in all types of education. In Article 5, the law establishes among its main objectives,

- Encouraging individuals and collectives, permanent and responsible, to engage in preserving the environmental balance and understanding the defence of environmental quality as a value inseparable from citizenship.
- The strengthening of citizenship, self-determination and solidarity as the foundation for the future of humanity (Brazil 1999).

As a democratic practice, environmental education is a preparation for the exercise of citizenship through active participation at individual and collective levels, given the socio-economic processes and political and cultural factors that influence it. The urban-industrial society and its current model of economic and technological development have had an increasing environmental impact but the perception of this varies in different ways between rich and poor.

People with low incomes have experienced directly the impacts of environmental problems. This is witnessed in their daily lives and expressed in water shortages, power shortages, the absence of safe living spaces, inadequate food supplies, among others. The reduction of social inequalities is essential for the achievement of full sustainability in all its dimensions, as was proposed by the International Conference on the Environment and Society: Education and Public Awareness Campaigns for Sustainability held at Thessaloniki, Greece, in 1977. Here, it was stated that the concept of education for sustainability should,

> …encompass not only the environment but also poverty, housing, health, food security, democracy, human rights and peace, resulting in their moral and ethical imperative in which traditional knowledge and cultural differences should be respected. Education and training for public awareness were considered pillars of sustainability together with legislation, economy and technology, involving the interaction of the collective effort of key sectors, rapid and radical changes in behaviour and lifestyles, as well as patterns of production and consumption (Pelicioni 2000: 52).

In a document analysing environmental education, Reigota (2003: 38) states that in Brazil many educators consider it as “an education policy that aims at citizen participation in finding solutions to environmental issues at local, regional and world levels”. According to the same author,

> Environmental education cannot lose sight of the complex challenges (political, ecological, social and economic) in the short, medium and long terms. In turn, the values of autonomy, citizenship and social justice are considered as basic principles of education…Autonomy characterises people who have a clear consciousness of its specificity in a given society. The idea of citizenship, based on political equality among all members of the nation, has enriched itself (p.39). By requiring the right to difference, which results from political participation, increasingly important social groups are organised being based on specific propositions and breaking the hegemony of the dominant discourse (indigenous, blacks, women, youth, elderly, etc.) (p.40) …The issue of social justice is a current issue in a society like Brazil, which is characterised by huge social, economic and cultural differences. This society will not become fair if there is not a fair distribution of social goods and the cultural conditions that produce them (p.41) …The various activities aimed at achieving a sustainable world will only be able to address the political and ecological aspects of our time if they include the requirement of justice (p.42).

The Emergence of the Sustainable Development Paradigm: The Brazilian Case

The term sustainable development first appeared in 1980 in *Global Strategy for Plant Conservation*, a publication associated with a conference organised by the International Union for Conservation of Nature and Natural Resources (IUCN), a Non-Governmental Organisation (NGO). It sketched out principles aimed at the elaboration of a Convention on Biological Diversity. However, the term sustainable development only became widespread in 1987 with the so-called Brundtland Report, sponsored by Norwegian Prime Minister Gro Brundtland who offered to hold meetings in cities around the world, including São Paulo, aimed at finding solutions to the environmental issues raised after the Stockholm Conference in 1972.

Debate about the sustainability of development dates back to the beginning of the 1970s, however, when the Polish economist Ignacy Sachs elaborated the concept of eco-development and discussed strategies for eco-development. Ignacy Sachs extended the concept of development to include not only economic variables, but also variables that incorporated political, cultural, social, ethical and environmental aspects. The basic principles of this vision of eco-development are summarised in the following quotation from Andrade (2000: 190),

(a) the satisfaction of the basic needs of all human beings in the present (synchronic solidarity);
(b) solidarity with future generations (diachronic solidarity);
(c) the participation of people involved in all development programmes;
(d) construction of a social system with guaranteed employment;
(e) social security and respect for other cultures;
(f) educational programme.

The declaration resulting from the 1974 Conference of the UNCTAD (United Nations Conference on Trade and Development) Programme and the United Nations Environment Programme (UNEP) as well as the report "What now?", presented at the end of 1975 by the Swedish Dag Hammarskjold Foundation, elaborated the concept of sustainable development proposed by Sachs without giving visibility to the term eco-development. Thus, the term sustainable development became widely available and was established by the 1987 Brundtland Report, which has as its first statement: "A global agenda for change". It is with this appeal that Brundtland, as chairman of the World Commission on Environment and Development (1991: 9–10), opens her foreword to the report, known in Brazil and round the world as *Our Common Future*.

Formally introducing the notion of sustainable development, the document affirms the difficulties in the convergence of the gamble on economic growth, the overcoming of poverty and the paying of attention to environmental limits,

> Humanity is able to make development sustainable to ensure that it meets the needs of the present without compromising the ability of future generations to meet theirs too. The concept of sustainable development has its limits – not absolute limits but limitations imposed by current technology and social organisation…Poverty is not an evil in itself, but, to be sustainable, development is necessary to meet the basic needs of everyone and give everyone the opportunity to realise their aspirations for a better life (p.10).

Although sustainable development requires comprehensive and global action, the concept was developed within the sphere of thought driven by an economic logic with reference to the existing dominant society. Thus, nature becomes a capital good in an ecology within a market economy.

The same thread is evident in the UN Conference on Environment and Development held in Rio de Janeiro in 1992 and, referred to as Eco–92 or the Rio92 Earth Summit. This is considered an important milestone in international efforts. It encouraged the advancement of the environmental debate in Brazil, promoting dialogue between social movements, non-governmental organisations (NGOs) and environmental movements. Many important institutions and organisations dedicated themselves to include discussions about environmental issues in their agendas, contributing to the broad public debate occurring at that time. This occurred in the so-called development NGOs, social movements, religious organisations of various orientations, the Catholic Church, universities, business sectors, among others.

The preparatory period for Eco–92 began in 1990 and the main consequence of this initial step was the focus on civil society. In May 1990, the Forum of NGOs and Social Movements in Brazil was created to monitor and participate in the conference. This involved a broad spectrum of agencies, such as those related to indigenous rights, women, residents' associations, youth groups, environmental organisations and environmentalists, trade unions, religious groups, development NGOs and social organisations of advisers. The forum brought together 1,300 organisations from 108 countries beyond the 1,180 activists of independent organisations. Called the Global Forum, it conducted in Flamengo Park a major event running in parallel with the official conference.

It was mainly from these discussions between different sectors engaged in the environmental and social struggle that arose one of the most important developments of this period: the notion that the problems being discussed were not exclusively social or environmental and could only be tackled if understood as the result of the convergence processes of society and the environment. This approach came to be known as the social, environmental approach and it helped to create a field of dialogue among many social movements.

Eco–92 was the first international meeting to have as its main objective the development of strategies seeking to halt and reverse the effects of environmental degradation. The conference focused on five important documents:

1. Rio Declaration or the Earth Charter,
2. Agenda 21,
3. Declaration of Forests,
4. Convention on Climate Change,
5. Convention on Biodiversity.

For this reason it is considered a milestone in international efforts towards sustainability.

The conference marked a time of great progress in negotiations between countries despite the problems that arose in closing the gap between conflicting views in the preparation and drafting of various documents. Thus the Biodiversity Convention was ratified only by countries with special biodiversity interests, like Brazil, but without the participation of countries possessing the technological means for research and industrial processing. For the Declaration of Forestry no negotiation was possible. However, both established themselves as political landmarks in the search for a better world. The Kyoto Protocol of 1997 was an unfolding of Eco–92 intended to meet the principles of the Convention on Climate Change. It was concluded in 1997 as a specific institutional arrangement for reducing emissions of gases from burning fossil fuels and the greenhouse effect. It was opened for signature in March 1998 and came into force in February 2005.

In 2002, the United Nations (UN) promoted the organisation of a new international conference to evaluate progress across the world toward sustainability over the 10 years since Eco–92. The World Summit on Sustainable Development, also known as Rio + 10, took place in Johannesburg, South Africa, with 193 countries and 58 international organisations participating. Here the decisions in the documents of Eco–92 were reaffirmed in the commitment to the interdependence between economic growth, social justice and environmental protection. In addition, major goals of poverty eradication, changing unsustainable patterns of production and consumption and protecting natural resources were established. However nothing was agreed as to the means of implementing policies designed to achieve the key objectives.

Resource transfers between countries as well as changes in the rules of international trade and finance were topics on which agreement could not be obtained. Current patterns of production and consumption were not included in discussions. In practice, they remained the prevalent reasons for defining narrow economic and financial goals at the expense of building a healthier world.

The historical setting in which the event occurred was quite different from the panorama of Eco–92, when the world was fresh from the ending of the Cold War. It had been scheduled to strengthen international solidarity in combating threats to security, compromising progress towards effective sustainable development.

The Brazilian Agenda 21, published in 2002, identified the following six priority themes for work and established strategies and specific actions for each of them:

1. Natural resources management
2. Sustainable agriculture
3. Sustainable cities
4. Infrastructure and regional integration
5. Reduction of social inequalities
6. Science and technology

The set of strategies and actions offered by Agenda drew various reflections, as described by Afonso (2006: 57),

> The first is that the notion of sustainability does not make sense if applied to a single sector of the economy such as agriculture, for example, or a specific system, such as cities. Thus, sustainable agriculture and sustainable cities are proposed that alone do not meet the principles of sustainability. The modification process must necessarily integrate the diversity of national life and, from this point of view, the Agenda proposes strategies that consider all aspects needed to build a new country.

As explained by Samyra Crespo, Secretary of Institutional and Environmental Citizenship, Ministry of Environment in Brazil.

> The Agenda 21 project was the agent for the mobilisation of hundreds of cities to gauge their problems in areas such as construction of homes, preservation of green areas, waste management, drinking water supply, conservation of energy and mobility. However, little was done to raise the welfare associated with the care of the environment. The result of years of actions shows that many municipalities were able to identify their challenges. But because of lack of investment, the projects were not implemented. I do not feel frustrated because there was learning (Flosi 2010: 12).

The Secretary asserted that to reverse the current unsustainable framework of cities and to ensure that the current problems of cities are not transferred to the future it is necessary to review the National Plan and Agenda 21. "We are far from achieving sustainable cities. The city of São Paulo, for example, while not solving the problem, that is urban mobility, is light years away from achieving sustainability" (Flosi 2010: 12).

Rubens Harry Born, one of the managers of the Brazilian Network of Local Agenda 21s (REBAL) Project agreed that Brazil failed to halt the processes of environmental degradation,

> Agenda 21 is no reference in the world. Most actions brought no results. Thus, the concept of sustainability comes as an alert for the next generations. The warning is of concern. We need urgently to follow the multiple paths existing, in the cities that were positive for achieving sustainability. The scenarios for the future of sustainable cities have already been plotted (Flosi 2010: 13).

Born defends his thesis based on "3HS" for successful planning and implementation of sustainable cities. They are: people aware, sustainable housing and healthy habitat. However, he argues that,

> The city of São Paulo is still in the opposite direction to 3HS. Urban mobility is an absurdity, consumption is disorderly and waste is uncontrolled. A good solution to urban problems in the city is public transport with quality. It is no use the government to create bike lanes for strolling if these bike lanes are not structured so that people can use them to get to work and study (Flosi 2010: 13).

He suggests that a sustainable city needs several other sustainable cities. He asserts, for example, that, "…it is no use being sustainable where I live if because of my sustainability an Indian has to lose his. Due to this need to find synergy between the cities surrounding the sustainability, there arises the need to find forms of consumption that respect all of this ethnic diversity, with its different cultures and different histories" (p.14).

Between 2007 and 2008, more than half the world population has been concentrated in cities, a fact unprecedented in human history. According to Golub,

> …urbanisation challenges our ability to produce public goods, especially education, culture, health and a healthy environment for all people, that are prerequisites for sustainable development to ensure the welfare of all persons and thus the expansion of individual liberties (2010: 6–7).

In Brazil, urbanisation occurred as a result of the expulsion of the rural population, in a process of accelerated growth of around 7–8% per annum. For such an increase it was impossible to provide the essential infrastructure and sanitation.

The city of São Paulo, for example, had 1.5 million inhabitants in 1950 but today this urban mass includes 11 million people, which, when combined with the population of surrounding municipalities, called Greater São Paulo, reaches 20 million people.

In Brazil, 84% of the total population of 195 million inhabitants is in the urban environment and there is a need for a corresponding capacity to manage its 5,564 municipalities. People who work in the city of São Paulo lose every day, on average, 2 h and 40 min of their lives in traffic. In São Paulo, the average speed of a car in traffic is 14 km/h. Roads constitute which 950 km^2 of impervious areas out of a total municipal area of 1,560 km^2. This means that when it rains the water has nowhere to drain, causing even greater inconvenience to traffic that is already chaotic. Within this framework is a Metro network of only 60 km. This network is small compared to, say, the 400 km of the Metro in Paris, a much smaller city than São Paulo.

The city of São Paulo is a city paralysed by water, the excess of cars and a very insecure public transportation system. There are many challenges in creating a sustainable city through, for example, deploying a network of good public transport, reducing carbon dioxide emissions, having 'green' buildings, monitoring the quality of air and water, managing waste, preserving vegetation and encouraging sustainable businesses. Of these challenges, trash is the biggest environmental tragedy in Brazilian cities. Crespo argues (2010: 14),

> Garbage is cursed. People throw out the trash, and think they've done their part...that is, after disposal, people and government do not care anymore. Have you seen any mayor elected because a landfill has been opened? There is no investment in waste collection and treatment because there is no social pressure.

According to the Ministry of Environment figures, almost 6,000 Brazilian municipalities reported that they piled up, per month, 150 t of waste. Only 4% of urban waste in Brazilian cities with more than 200,000 inhabitants is recycled. Compounding this situation is junk mail, or e-waste, which takes considerable space on the platforms of discussion about recycling. Annually, 40 million tons of e-waste in the world are discarded. Among the emerging countries, Brazil occupies the first position, showing the urgent need for initiatives that promote changes in the scenario shown above.

Faced with these problems, an initiative that is gaining prominence in the city of São Paulo is the Cooperative Production, Recovery and Reuse, Recycling and Solid Waste Electric and Electronic Marketing. This encourages the sending of products sold by manufacturers, as well as any type of e-garbage, to the cooperative. Upon receipt of the material collected by the cooperative's truck, material is sorted and recycled. Dealing with e-waste is critical because this kind of trash has a high degree of toxicity and a high potential for contamination.

For Brazil, the major question that arises is how to translate the theoretical principles relating to social and economic dynamics into practice. Brazilian society is not moving in the direction proposed by Agenda 21. Instead, all sorts of imbalances can still be found aggravated by water and air pollution, pressure on natural resources, deforestation, income inequality and unemployment.

Environmental Education

During Eco–92, arising from an unofficial discussion forum held between representatives of Non-Governmental Organisations (NGOs) and civil society, a Treaty on Environmental Education for Sustainable Societies and Global Responsibility was signed. This document stressed the guiding principles of environmental education previously established in international environmental education. The treaty emphasised that, "Environmental education is not neutral, but ideological, it is a political act, based on values for transformation" (Department of the Environment of São Paulo, 1994: 8).

In 1997, two very significant events marked the evolution of environmental education in Brazil: the Fourth Forum for Environmental Education and the First Meeting of the Network of Environmental Educators, promoted in the city of Guarapari, in Espírito Santo along with the First Conference on Environmental Education held in Brasilia. These meetings resulted in the drafting of a document concerning the ideas for and Brazilians' contributions to environmental education, and was directed at the International Conference on Environment and Society: Education and Public Awareness for Sustainability, sponsored by UNESCO in December 1997 at Thessaloniki, Greece.

The Brasilia Conference reaffirmed the pertinence for Thessaloniki of the resolutions set out in several previous conferences sponsored by the UN. It called attention to critical issues such as the need to invest in teacher training, the lack of teaching materials, the lack of national policy and clear strategies for the implementation of environmental education, the lack of evaluation of the actions taken and the difficulties of promoting changes in values through educational practices. The concept of sustainability was extended to encompass not only the environment but also poverty, housing, health, climate, security, democracy, human rights and peace, resulting in a moral and ethical condition in which cultural diversity and traditional knowledge are respected.

Within this new concept of sustainability, actions for environmental education are diverse but, at the same time, contradictory. In São Paulo, for example, less than 1% of household waste is recycled where experts say the rate could reach 30%. This situation resulted in a case that is still being decided by the Brazilian courts asking, too, for the deployment of recycled materials to landfills and promoting cooperatives for recyclable materials. However, in the same period of the lawsuit, the city of Curitiba, in Parana State, received the Globe Sustainable City Award 2010 as a model city for sustainability in the world. The Prize was awarded by the Globe Forum, a Swedish organisation that brings together entrepreneurs concerned with global sustainability. According to the organisers, Curitiba was chosen because it showed maturity in understanding the importance of sustainability. The organisation also praised the integration of environmental spheres into the intellectual, social and economic life of the city. The main programme presented by Curitiba was the Biocidade (Biocity) programme that conditioned all actions of the municipality within environmental issues. Because of this, the city has now, on average, more than 50 m^2 of green area per inhabitant.

In June 2010 an International Conference at Luziânia, state of Goiás, brought together 600 young people aged 12–15 from some 50 countries to discuss the social and environmental problems of the planet. This International Conference, "Let's Take Care of the Planet", has promoted intercultural dialogues as well as playful and practical workshops to generate actions. A document was drafted called a charter of responsibilities where these young people assumed collective responsibility and suggested changes that must be brought about globally and locally. This conference a part of the Decade of Education for Sustainable Development (2005–2014), defined by the UN for which Brazil has already held two national conferences entitled "Let's take care of Brazil". The first held in 2003 and a second in 2009 brought the participation of about 13 million people from 20,000 schools nationwide that became centres of debate.

The contradictions surrounding the concept of sustainability can also be viewed, for example, in the outcome of an event that occurred between October 9th and 11th 2010. The SWU – Music and Arts Festival, focused on sustainability and brought together about 150,000 people in the city of Itu, São Paulo. Despite presenting a Public Commitment to an Action Plan for Sustainability and bringing together experts, scholars, businessmen and representatives of non-governmental organisations and other initiatives relating to Brazil and the world, to discuss with the audience some of the key issues of sustainability, the event generated approximately 3 t of garbage that could have been stored in dumps for recyclable materials located throughout the site. But this garbage was left on the floor of the event showing that talking about environmental education and environmental practice do not always converge.

Work still needs to be undertaken, and a major exhibition showing that the planet has already surpassed some of its own limits on account of the impacts it has suffered over time and how little time there is to correct it will open the Global Forum for Sustainability in 2011 in Rio de Janeiro. This is an opportunity to stimulate debates in public, private and civil society in order to make Rio de Janeiro, which will host major world events, a sustainable city. The idea began on October 7th and 8th 2010 when about 100 people from around the world gathered to discuss how it should happen. The first activity was to propose actions to influence the River 20 – International Conference of the United Nations, marking in 2012 20 years since Eco–92. The main point of discussion was finding ways to ensure that any actions taken in Rio20 will not be doomed to failure.

The UN Under-Secretary General and Executive Director of the Institute for Training and Research of UNITAR (United Nations Institute for Training and Research), Carlos Lopes, who was a leader of Eco–92 and is currently organising the Global Forum for Sustainability, describes the landscape favouring sustainability as being in a state of paralysis,

> The big question being discussed now is how we will have a global governance, taking into account the fact that one does not reach agreements on key issues on the planet. Organisations are structured in the form of silos, the conventions are stopped. There is no shortage of ideas, proponents, actors, but a lack of governmental engagement against those ideas (Gonzalez 2010: 6).

According to Lopes, Brazil's role in global negotiations for the environment is important because the country is one of the principal actors in the area of international trade and has relatively cleaner energy sources and greater biodiversity than many other parts of the world. Moreover, "...as seen in the last elections in the country and the world, the environment has entered the domestic agenda" (*idem.*).

Organisation of This Book

We have divided the book into three parts. Part 1 introduces key ideas that underpin much thinking and practice about sustainability throughout South America. Through a focus on a small number of countries it is possible to highlight some of the commonalities and differences in environmental and environmental education policies that can be explained by particular, national circumstances.

In Chap. 2, Boschetti identifies some of the major forces that impact on all of the nations of South America. Modernisation, industrialisation, globalisation and urbanisation have powerful meanings in a South American context. They demand a rethinking of the importance of the natural environment and the relationships between people and their environments, local, national and global. This rethinking has ethical and moral dimensions, drawing upon considerations of social justice, autonomy and citizenship, as highlighted in our introductory chapter. These philosophical, social, economic and political elements recur throughout the book and must be thought of, simultaneously, in national, global and international contexts. They have a fundamental role in environmental education for sustainable development and the essential challenge is engaging young people with these key ideas while translating them into practical activities.

In the subsequent four chapters in Part 1 the focus is on particular nations. Chile is undergoing curriculum reform and in Chap. 3 there is a detailed examination of the role of teaching about sustainable development within the school subject geography. A need had been identified to modernise this subject and a number of themes were considered important in this regard, including the spatial pattern of human occupation of territory, the interrelationships between society and nature, an holistic conception of the planet as the home of human beings, the spatial impacts of the globalisation process, global warming, migratory flows, fast urbanisation, the location of transnational companies, relationships between Chile and its region and the global economy, foreign treaties and Chile's impact on the geographic space. A Spatial Geographic Progress Map has been designed to monitor learning using a number of levels. This chapter identifies the strength of the traditional subject-based curriculum in schools and both its strengths and weaknesses in terms of encompassing education for sustainable development.

In Chap. 4 attention switches to Bolivia. Drawing on Norbert Elias's figurational theory and Leonardo Boff's theory of care the authors identify important social and ethical ideas that contribute to an understanding of education for sustainable development. They argue that the ethics of care, of compassion, responsibility and cooperation offers a significant challenge for future generations. They acknowledge that essential social changes will take a long time and such changes are frequently unplanned and unintended. Social change requires changes to individuals and to their society that are not separate but linked within a social structure. The evolution of the Bolivian educational system is explored and the educational implications of contemporary environmental problems are discussed, particularly issues relating to the cultivation of and trade in coca and the migration of Bolivians to Brazil.

Argentina is the focus for Chap. 5 and, as in Chap. 3, the school subject geography has been selected to highlight some aspects of educational reform that are associated with education for sustainable development. The formulation of a national strategy for the environment was followed by a national strategy for education for sustainable development. However, proposals for introducing cross-curricular arrangements have been largely unsuccessful because of resource problems and also difficulties in achieving an appropriate level of coordination between responsible national and local agencies.

Finally, in Part 1, Chap. 6 moves away from the curriculum and the authors have provided a review of broader educational reforms in Venezuela. Important historical trends are outlined leading to the current attempts to give a new meaning to education that encompasses principles of inclusion, multiculturalism, multiethnicism, and plurilingualism with a strong qualitative focus based on humanistic principles. This chapter pays less attention to the natural environment and emphasises more the social component in the society-economy-environment triad that lies at the heart of sustainability.

Part 2 is devoted to a detailed consideration of education for sustainable development in Brazil. As emphasised in the introduction to this chapter, Brazil is the largest nation in South America in terms of both area and population size. It epitomises many of the historical forces that contribute to an understanding of contemporary economic, social, cultural, political and educational aspects of South America.

In Chap. 7, the focus is on cross-curricularity and the environment as one of a number of nationally identified cross-curricular themes. This is an essential part of the reform proposals associated with the National Curriculum Parameters introduced by law in the 1990s. Cross-curricularity extends beyond a definition of knowledge across subject boundaries to include teaching approaches focused on active student participation. The study of the environment is perceived as an important component of citizenship education. This is evident in the work of the Schools for Tomorrow Project which is described in detail and supported by a small scale empirical study.

There is evidence drawn from another small scale study in Chap. 8. This chapter opens with a review of the principal stages in the development of the Brazilian educational system. This review concludes with the attempt to introduce the cross-curricular themes discussed in the previous chapter. The view of the current arrangements for introducing education for sustainable development is largely pessimistic. It is argued that the sustainability concept is organic and dependent on the harmony between those social, economic, cultural and environmental diversities that involve society in a general way. It is the absence of this harmony that leads to pessimism. For the poorest students school is seen less as a force for socialisation and more as a source of economic relief. They care little for their environment and much more for finding the means to survive. The more affluent students seemed to be more worried about the environment and the improvement of their social condition. However, their wishes for social mobility, the acquisition of new technologies and the search for comfort does not allow them to notice the environmental impacts caused by their life style. Theirs was a life of comfort and social climbing. There are serious challenges facing advocates of education for sustainable development in Brazil where, the author argues, where there is evidence of a lawless society where, families do not provide the ethical fundamentals, schools do not educate, churches do not indoctrinate, health care does not cure, the banks do not work, the political parties do not politicise, the police do not police and the government does not govern. These, it is claimed are the usual consequences of the negative results of education and official statistical data are used to demonstrate the substantial weaknesses in the education system.

The next two chapters consider education for sustainable development at the outset and the termination of formal education. They are concerned with early childhood education, on the one hand and university education, on the other. The bulk of Chap. 9 is taken up with an account of the project *My Neighbourhood, My Land, My Treasure* that was undertaken with elementary school students aged 8–12 years. This project was based on a school-community partnership and was driven by the need to see schools as institutions concerned less with narrow vocational training and more with inculcating a range of broader values. In particular the project leaders wanted the students to address the causes and consequences of excessive consumption of material goods. Improved human relationships and greater autonomy of individuals were regarded as important outcomes. Contrasting with the emphasis on the young learners' immediate communities, the project described in Chap. 10 covers a much wider geographical area. The multi-campus Federal University of Tocantins was established in 2003 in the Amazon region with a mission to engage in teaching, research and extension activities that would contribute to both regional economic development and improvements in the quality of life in the region. The authors explore the theories underpinning this innovation and discuss the many challenges that university planners face as they seek to fulfil their goals of achieving regional sustainability.

In Part 3, the focus is upon a selection of projects that highlight future trends and challenges. In Chap. 11, the authors describe an international collaborative project in the field of industrial design. Students in Italy and Brazil were set the task of designing household objects using raw materials derived from industrial waste and employing technologies and production processes that were commonplace for artisans. The project participants developed a model that promoted social inclusion of people from poor communities who worked with the reuse of recyclable waste. The project illustrates the important society-economy-environment triad that lies at the heart of the sustainability concept.

If Chap. 11 adopts a broadly utilitarian and practical approach to project development, by contrast Chap. 12 is much more concerned with whole school cultural transformation with particular reference to a new moral and ethical order. Sustainability, is viewed as as a multidimensional problem demanding of schools responsible reflection with regard to future generations and requiring them to educate morally autonomous people. Such individuals would be capable of critically analysing political attitudes, values, rules, and laws that determine collective social life in an interconnected environment necessary for life on the planet. While acknowledging that students have their own personal ethical convictions, influenced by religion, life history, and experiences which make up their uniqueness, it is argued that a collective stance on the degradation of the earth's ecosystem is urgent. Profound changes are needed in school organisation and pedagogy to meet these ethical challenges.

Challenges of a different order are the subject of Chap. 13. The work of a small scale, local information technology project titled *Vozes que Ecoam* (Echoing Voices) is described. This sought to promote an alternative form of digital inclusion in an out-of-school environment and aimed at rebuilding the identities of the participants and their communities. The project activities introduced a critical pedagogy based on visual, verbal and digital literacies and it is argued that this pedagogy offered both an alternative curriculum and the chance to rethink the needs of local communities. Through the project it is possible to see a convergence of a sustainability discourse and a pedagogical discourse. Digital technologies are powerful tools capable of both controlling and emancipating those who are able to access or purchase them. There is no doubting their importance in schooling for sustainability but there is a need to monitor their influence on young learners.

Just as our introductory chapter identified three universal themes in sustainability discourses (autonomy, citizenship and social justice) the author of the final chapter returns to globalisation and urbanization as two universal themes. Both of these are significant in the recent histories of South American nations and the links between them and schooling are discussed.

References

Afonso, C M. 2006. *Sustentabilidade: Caminho ou Utopia*. São Paulo: Annablume.

Andrade, J B T. 2000. *Desenvolvimento Sustentado e Meio Ambiente*. Brasília: Universidade de Brasília (DF).

Brazil. 1981. *Lei n.6938, de 31 de agosto de 1981. Dispõe sobre a política nacional do meio ambiente, seus fins e mecanismos de formulação e aplicação e dá outras providências*. Brasilia: Diário Oficial da República do Brasil; 2 set. Seção I: 16509.

Brazil. 1988. *Constituição da República Federativa do Brasil*. Brasilia: Ministério da Educação.

Brazil. 1999. *Lei n. 9795, de 27 de abril de, 1999. Dispõe sobre a educação ambiental, institui a política nacional de educação ambiental e dá outras providências*. Brasilia: Diário Oficial da República do Brasil.

Carvalho, I. 2001. Qual educação ambiental? Agroecologia e Desenvolvimento Sustentável. Porto Alegre 2(2): 43–51. April/June.

Flosi, C. 2010. Horizonte Sustentável. *Cadernos SESC de Cidadania*: *Dia do Meio Ambiente* – June 5th. I(11): 8–15.

Golub, S. 2010. *Saturação das Metrópolis*. São Paulo: Le Monde Diplomatic Brasil.

Gonzalez, A. 2010. Rio Fera Fórum pela Sustentabilidade. *O Globo – Razão Social*, October 19th. No. 105(6): 4–7.

Pelicioni, M C E. 2000. *Educação em Saúde e Educação Ambiental: Estratégias da Construção da Escola Promotora Da Saúde*. São Paulo: Faculdade de Saúde Pública da USP – Tese de livre docência.

Reigota, M A. 2003. Educação ambiental: uma busca da autonomia, da cidadani e da justiça social. o caso da América Latina. In *Educação Ambiental: seis proposições para agimos como cidadãos. Aliança para um mundo responsável, plural e solidário*, ed. Y Zialca, C Souchon, and H Robichon. São Paulo: Instituto Pólis.

SMESP (Secretaria do Meio Ambiente do Estado de São Paulo – Coordenadoria de Educação Ambiental). 1994. *Educação Ambiental e Desenvolvimento: Documentos Oficiais*. São Paulo: Série Documentos.

Chapter 2
Sustainability and Educational Actions

Vânia Regina Boschetti

> Alienation entails not only the object of work, the work process, the creative and transformative activities that define human species and the workers related to the others, but also, and always together, the relationships between humankind and internal and external nature.
>
> Foster (2005)

The thoughts expressed in this quotation, derived from Karl Marx in his *Economical and Philosophical Manuscripts*, introduce the theoretical elements on which we shall focus in this chapter. For Marx, nature was considered an extension of the human body, its inorganic expression. Without nature, there was no humankind. Amplifying this, Foster (2005: 107–108) asserted, men "produce their own history producing their means of subsistence mostly. Nature becomes practical to human kind".

To continue this reasoning, we can affirm that the human trajectory on the planet is characterised by the occupation and transformation of natural spaces. The process of industrialisation originating in the seventeenth and eighteenth centuries was distinguished by disrespect for nature and the exploitation not only of the land and its resources but also of other human beings. Increasing their basic needs to a superfluous level without making any kind of valuation of the possibility of alternative ways of meeting them, people failed to assess the efforts needed to obtain them. Referring to the reflections of Pitano and Noal (2009: 287) on this matter, it is possible to affirm that, "the territories and the resources inside them are seen like 'space to be taken'...and the people are considered as instruments of this conquest too".

By combining these two factors, industrialisation and natural resources exploitation, and adding to them the idea of development, we get the bases of capitalism. People used technical and technological improvements for exploitation and for

V.R. Boschetti (✉)
University of Sorocaba (UNISO/SP), Sorocaba, São Paulo, Brazil
email: vania.boschetti@prof.uniso.br

M.L. de Amorim Soares and L. Petarnella (eds.), *Schooling for Sustainable Development in South America*, Schooling for Sustainable Development 2,
DOI 10.1007/978-94-007-1754-1_2, © Springer Science+Business Media B.V. 2011

generating pollution that have modified the living space. The gradual intervention of people in the natural environment modified it, causing an environmental lack of equilibrium and creating a "new time, at work, in the interchange, at home. The social times tend to surpass and go against the natural times", which multiply and deepen the human action over space (Santos 2002: 236).

Santos points out that the advance in techniques was substantially increased after the middle of the twentieth century. The deliberations of the global market and the demands of globalisation fragment and bind the current world producing global spaces that rule, give the orders (hegemonic) and require others to execute those orders. The world is no longer thought of as a totality but as the sum of spaces taken separately. He argues that one of the globalised world's characteristics is the demand for fluidity in the circulation of ideas, messages, products and situations that lead to competitiveness. Because of this, all kind of objects and places are created – from canals to intelligent districts, from racetracks to industries, from spaces for leisure to spaces for confinement. Globalisation has been linked to the idea of nets that operate at world speed (the fluidity referred to by Santos) generating a whole system of dependency and a new world architecture sustained by the universality of a single technical system, where there is no global space, but spaces of globalisation.

To reinforce these ideas of Santos, Bauman (2003: 89) explains that globalisation is a process that is not followed equally by all institutions; deep down it is linked with the unequal development of economics, politics and culture. Power becomes extra-territorial because it stays incorporated in the world circulation of capital and information. The political institutions, on the other hand, continue to be local and weak for not being capable anymore of raising resources necessary to execute independent policies. National governments have no other option but, to abdicate control of economic and cultural processes and deliver them to 'market forces'.

Returning to Santos, we notice that globalisation represents brutal value changes. Before globalisation took hold, the valorisation processes were relatively slow. Today there is a changing process that compromises personal reorganisation and integration of the historical subjects because this globalisation does not 'think' about people. There is no space for emotions, the link that should continue to unite people.

A proposition emerges: the economic question is directly linked to the environmental problem and to life on our planet. Once the economy discovers its main resources for the production of goods, the exploitation of natural resources follows. Immediate consequences are global warming, deforestation of the Amazon Basin, extinction of wild animal species, the melting of ice sheets at the poles, pollution at different levels and weather changes that put the continuity of life on the planet in danger.

When Francis Bacon, the empiricist English philosopher of the sixteenth century, affirmed that nature should be exploited to benefit human interests, he probably did not think about the outcomes of this. As has already been affirmed, the exploitation of natural resources over the centuries, guided by uncontrolled attitudes, has been compromising on a large scale the natural reserves and their preservation. Many species have simply ceased to exist and many others are about to become extinct. The planet does not have the means to sustain the usual standard of living, especially in the North American and European countries.

Why is this problem so difficult to resolve? In part it is because the apparent nonsense is just a superficial aspect of more complex questions. The logic of economics is more important than the interests of the collectivity in a context where profit is sought more and more in nonsensical ways. The notions of development, underdevelopment and modernisation lie at the heart of the system of ideas that has guided international politics in the West. The development paradigm is derived from American society and it reproduces as closely as possible, the 'American way of life' with a prevailing idea of unlimited development of productive power, intensive use of raw materials and the exploitation of non-renewable sources of energy. For the Latin American countries to move to this condition it is necessary to achieve economic growth, industrialisation and urbanisation, even though these demand sacrifices like a growing debt, marginalisation and devaluation of popular knowledge and exploitation of the labour force and natural resources.

"Misery forces free men to do many selfish things", asserted the great classical Greek orator, Demóstenes in *Orationes*. In the foreground of the picture painted by theoreticians addressing contemporary problems, we notice that it is not just misery but it is also greed, the search for money and power, so typical of the abundance society, which lead to an even stronger way of engaging in mean, unfair and irresponsible practices.

The global dynamics, as Gonçalves (2003) points out, spread a youth consumer culture which is compounded by an infinite number of items that are constantly offered to these consumers, including music, fashion, entertainment and leisure items. The young, bombarded with symbols of power, success and group identity, intensify their need for their 'objects of desire' and the impossibility of having them generates various frustrations and promotes a variety of ways of obtaining compensation.

Reigota (1999a: 65) points out that those paradigms and models of hegemonic groups, created by the consumption and obtaining of goods, brings closer together chronologically defined sectors in the same population who are distant from each other in terms of resources stratification: the rich, the starving and the intermediate,

> …that occurs not only with the so called developed countries, but also in many places from the countries considered to be underdeveloped. This life style has the consumer as its reason to be. An enormous quantity of perishable, superfluous and disposable goods is consumed, besides natural non-renewable resources, and oil pollution is derived.

Obviously, despite the strength of economic power and the social representations of it, the consequences of its expansion begin to appear in the everyday lives of people in various countries. It has become obvious that the continuity of an economic model based on the exploitation of nature and the predatory use of natural resources is irrational. The first manifestations of this were those social movements, of which the most influential was, possibly, the counter-culture movement, identified with the internationally troubled year of 1968. Because of these movements, ecological questions took on a world dimension with the support of NGOs, the mass media, pacifist and anti-nuclear groups, international organisations, social educational agencies and civil associations directed at environmental action and management. According to data provided by Scotto, Carvalho and Guimarães (2009), by the end of 1974 more than 60 countries had national environmental organisations, a figure that reached 100 by the end of the decade. The NGOs increased from 2,500 organisations in 1972 to 15,000 in 1981. It must be pointed out that in that period there were several significant international events relating to the environment involving a wide range of social and governmental groups. As examples, we can mention the Meeting of the Club of Rome and the United Nations Conference for the Human Environment held in Stockholm, both in 1972.

It is important to notice, as affirmed by Jimenez and Terceiro (2009), that the ecological movement, the green movement, environmentalism and environmental education appear as outcomes of a larger social mobilisation that included NGOs, sympathisers, activists and political militants. Their action offered "guides for environmental protection and, in this context, it is a cause considered to be closely linked to social problems" (p.304).

In 1975, under the auspices of UNESCO, some of the earliest environmental education conceptions were communicated through the Belgrade Letter. Besides those issues of a more general nature, we have to consider much more precise aspects of the political-economic structure that define the current world scene: the changing of the geostrategic poles represented before by the socialist and capitalist sides to defining a new reality delineated on one side by the wealth and development of the northern countries and, on the other side, by the poverty and underdevelopment of the southern countries. The focus on the environmental crisis has become part of the discussion of global priorities marking it as a time of heated discussions about the established development process, economic policies and their effects on the planet (Reigota 1999a).

From the atmosphere of criticism with proposals for limiting development emerged the concept of eco-development, a precursor of the idea of sustainable development. The concept of eco-development, according to Sachs (1986), is a creative process that changes the environment through the use of careful techniques, respecting the potential of the environment without waste and using the environment to serve real needs, meaning that it requires an educational and organised action plan. This conception forces a confrontation between the defenders of 'zero growth' and defenders of 'nonsense growth', with the proposition of ecologically directed and, therefore, careful development. The analysis of the problems gained a qualitative dimension.

It is from the end of the twentieth century, more precisely in the last decade, that discussing environmental matters began to be more common, for many reasons: the establishment of the limits of development for the solution of global problems; the seriousness of the consequences that nuclear accidents like Chernobyl highlighted; the practice of nuclear testing made in several countries; discharges from the chemical and pesticide industries into the soil and into spring waters; the emission of toxic gases; the use of agrochemicals that reduce the quality of life; and the different kinds of intense and aggressive cultivation that have brought risks and dangers to the whole planet and its diversity of life. The United Nations Conference for the Environment and Development, held in Rio de Janeiro in 1992, helped to popularise the ecology movement.

It is in the late 1960s that the first activities in the ecological social debate have their roots, as well as marking the genesis of the sustainable development concept. This concept recognises the inequality evident all over the world and the threat that poverty brings to a social and environmentally balanced future for all of us. The sustainable development concept appears in the Brundtland Report *Our Common Future* as an outcome of the work of the "Comissão Mundial sobre Meio Ambiente e Desenvolvimento" (Report of the World Commission on Environment and Development: Our Common Future 1998) (CMMAD), published in 1987 in England and the United States (Scotto et al. 2009: 9).

The notion of sustainability is linked to a practical logic, being concerned more with desired social effects than with the scientific knowledge field in which analytical concepts are constructed to explain reality (Acselrad 2001). It means, from a specific point of view, integrated development with nature as well as all humankind. In a more comprehensive way, advocates of sustainability believe in spreading resources and in social and corporate responsibility, in dialogue and in the constant up-grading of the literature about the best sustainability practices. It implies a necessary interrelationship between social justice, quality of life, environmental equilibrium and a rupture with development standards. As a new requirement, it must stimulate recognition of responsibilities as long as aspects like equality, social justice and ethics prevail over living beings.

The term 'sustain' has a precise meaning: it refers to support, sustenance and the ability to avoid collapse. According to Camargo (2003), the adjective 'sustainable' refers to that what can be supported. From this concept is derived the concept of 'sustainable development'. Sustainable development refers to the ability to support the relationships between resources and production. It tends to view development from the perspective of capitalist standards that in principle and in actions conflict with the notion of sustainability. To resume this concept,

> Sustainable development would translate the new guidance paradigm about the relationship between society and economy with nature trying to deal with current needs, by improving the quality of life, economic progress, social justice and environmental quality without compromising the chances of total survival of future generations because of the destruction of the environment (Jimenez and Terceiro 2009: 307).

This new discussion has promoted a diversified and not always unanimous list of positions, contestations and classifications that have involved the major environmental matters linked to consumption, globalisation, rights, territory, material resources, economic production, and others. The solutions are also not unanimous. It is really very difficult to confront the power of the production relationships in capitalism because of the number of restrictions that are put up against the change possibilities of the pro-social sustainability technical options.

The complexity of these matters, the extreme fragility of the situations that polarise such different positions from companies and science, the strength and quality of life of the population, profitability and the end of natural resources, makes homogeneous possibilities and actions much more difficult. Sometimes, it appears to be like a mine-field of conflicting interests. Sustainability faces significant challenges to combat the already identified problems. The report of the Brundtland Commission (1987) establishes that,

> The concept of sustainable development has, of course, limitations – but not absolute limits but limitations imposed by the current technology and the biosphere absorbs the effects of human activity. Technology as well as social organisation can be directed and improved to create a new age of economic development...Poverty is no more inevitable...but for sustainable development to exist we have to deal with everyone's basic needs and give all the opportunity to achieve their wishes for better lives (CMMAD 1988: 10).

Starting from the sustainability concept of equilibrium, limits and biological ability and the possibility to improve life conditions on Earth it is thought that formal education in all its levels offers a possibility and a path. *Par excellence,* schools and colleges can be spaces for conscious thinking about and acting on the quality of life and the health of, and in, the planet. By initiating discussions inside educational institutions, it is possible to analyse environmental issues within the scope of the students' and teachers' daily lives, and to insert these issues into routine activities and practices that can lead to intellectual and behavioural changes.

There is a need for a more elaborated understanding of environmental matters that goes beyond common sense, leading to a clearer idea of the preservation of a person's own true existence. Once the approach has been fitted into the educational universe, popular culture and even religions, it works like a catalyst for different groups sharing the same aims with regard to these concepts and questions. We need, however, by taking the environmental field as a reference, to advance beyond so-called common sense and the trivial information broadcast by the media about scare stories and those issues that hold the public attention only momentarily. Jimenez and Terceiro have asserted,

> ...politics and the established thought about education for sustainability pointed out here implies analogously untangling the relations between work, education and the social reproduction complex, at what level and through what attitudes such paradigms and policies would be in the way of the demands of the system facing the current crisis (2009: 299).

With regard to the environment, there are many educational experiences and projects that have become part of the agendas of many school teachers. They have been incorporated with other existing conventional practices such as garbage collection and the careful and rational use of water and energy. New ways of thinking and acting must be introduced into the daily life of schools. As an example, curriculum development from the beginning of school age goes deeper into the questions and promotes the organisation of activities involving and linking different subjects by adopting effective practices. There are activities that allow teachers and their students to reflect on aspects of sustainability. These include aspects of human relationships in an environment where development is balanced, demanding active citizenship and an organisational structure that leads to a consciousness of rights and responsibilities and a clear understanding of the precise meanings of freedom, equality and human dignity.

The methods used by teachers in schools need to link learning situations with attitudes towards sustainability through the selection of appropriate lesson content and through the everyday dynamics incorporated into personal life and social life. Reigota (1999b) refers to this as making sustainability something you learn *with* somebody and not *from* somebody. The educational process through its extensive reach establishes a pathway where it is possible for new links to be made between people, nations and cultures, especially when talking about solving problems that do not have any bounds.

Examples of Courses

To indicate how, in educational institutions, aspects of sustainable development and sustainability can be incorporated, we can review briefly what has been undertaken in a number of Brazilian colleges where specific courses are offered that focus on contemporary realities including the production/resources breakdown, the deteriorating quality of life and the production of academic knowledge in some subjects.

What has been observed in these courses is that the institutions have tried to attract groups who are sensitive to environmental issues or who are linked to them. Generally, the academic propositions of these courses show a necessary technical quality and commitment to a 'business vision', with the production of social relations distributed under the rules of property and/or appropriation. To illustrate this, the following is a brief summary based on statements collected from course documents and institutional websites:

1. **Environmental Management**: this course is offered in two different ways, a technological course and a college degree course. Aimed at the education of the environmental manager, it is a professional course including basic knowledge of different fields in the environmental sciences. The course offers a specialised education in economics and sociology, biology and natural resources management. The academic contents are directed to flora and fauna, nature preservation, remote sensing and ecology, together with the development of those skills needed to manage environmental activities and 'knowhow' relevant for working to resolve problems related to soil pollution, water pollution, air pollution and the use of energy resources.

   Professional education in Environmental Management is linked to the elaboration and management of projects, especially those related to the preparation of studies and reports on environmental impacts. This activity requires close collaboration with many market sectors: private companies, NGOs, public institutions, institutions with production mechanisms that are in the front line of the environmental and sustainability problems, such as the chemical industry, mining companies, the iron and steel industry and the cellulose industry.
2. **Biotechnology**: as a field of knowledge, biotechnology brings together methods that can be applied to the activities that link complex organisms and their derivatives, employing many recent technological innovations. The use of the term biotechnology was established in the twentieth century and it focuses on biological

processes that contribute to the production of materials and substances for individual, medical and pharmaceutical uses. As an academic activity it can be extended to include the study of such environmentally related topics as human relations, the consumer and the demands of the market according to the economy and society.

Starting from the first uses of biotechnology, still rural and linked to fermented foodstuffs production, the field became more and more complex and, with the expansion of trade relations and the gradual increase in knowledge, it began to offer its services to more precise applications, including genetic knowledge of DNA, technology to produce penicillin on a large scale, and genetic engineering (animal or vegetable) support. Especially in Brazil, biotechnology is responsible for the improvement of processes that use sugar cane to produce ethyl alcohol and for sparing the environment from fossil combustion emissions, mostly by vehicles.

3. **Environmental Journalism**: in recent years, environmental journalism has been an active part of the communication media, as ecological issues have been addressed more and more in newspapers and other media. Facts, processes, studies and researches related to environmental conservation, diversity and sustainability have appeared as headlines and have been receiving the full attention of the media.

   Journalism, in Brazil, relates to traditional mass media, like magazines, and newspapers, as well as to smaller, more specialised media. However, it does not depend on immediate ecological questions. Scientific expansion, the less conventional practices in the fertilisation area, human reproduction, changes in the genetic code, deforestation, genetically modified crops, risks to diversity and national sovereignty, food security, all have been fertile fields for this kind of journalism. Environmental Journalism also refers to urban garbage production, to industrial pollution, inappropriate or condemned buildings, while trying to adopt a critical attitude.

What can be noticed from institutional statements is that colleges have already, or could be, extended beyond their original curricular areas: Biological Sciences diversified their academic provision, attracting applicants to an education that includes Health, Communication, etc. An opposite idea that would be more logical is some kind of specialisation that focuses on the major political structures defined by hegemonic institutions and changes in production relationships.

Recently in Brazil there has been a tendency for institutions to promote educational research studies in the sustainability area, especially with regard to the large number of PhD courses in Education and Environmental Education. It is important to point out that,

> …the approval of the environmental education national policy as well as the creation of its administrative organ permitted a better expression for the educational component of the growing environmental movement awakened since Rio 92. Another factor that must be pointed out is that, since 1997, the Ministério da Educação, through the Parâmetros Curriculares Nacionais – PCNs, defines the environmental themes, more specifically Environment and Health, as a cross-curricular theme at all levels of education (Gaudiano and Lorenzetti 2009: 193).

Educational activity occurs in social and cultural contexts. Considering the present problems and highlighting the environmental ones, academic education at any level should aim at sustainability. The statements of Gadotti (2000: 79) are especially illustrative,

> ...the success of the ecological attack does not depend only on conscious actions about, for example, the necessity of reforesting, cleaning rivers, making selective collection and recycling garbage. It is also necessary to solve the social problems, to relate the environmental questions to the social questions. In a picture like this, the sustainable development idea has a formidable educative component: the preservation of the environment depends on an ecological consciousness and the consciousness formation depends on education.

Going back to the previously mentioned development of research studies, Pato et al. (2009) mapped the trends in the subject matter in studies undertaken at ANPED (Associação Nacional de Pós-Graduação e Pesquisa em Educação) in two periods: 2003–2004 and 2005–2007. The studies were chronologically grouped to analyse the evolution of reflection and thought. The authors "confirmed Environmental Education as an epistemological scene characterised by theme plurality and by the convergence of multiple branches of education linked to the changes in the human-nature relationship" (2009: 217).

It is possible, despite the diversity, to notice that the studies emphasise the influence of reality in a way that modifies the relationship that people must have with themselves and with other people. Pato, Sá and Catalão (p. 232) point to, "the vocation of the Environmental Education researches for changing reality to more sustainable relations to the people, the groups, the society and the planet in an action-reflection-action permanent process".

The academic studies were grouped into three categories: topics, theory and methodology. After this, an analysis of the pieces of work was made by locality in each category from the evolution in the topic line of ANPED; then the research behaviour was analysed, as well as its location in the methods of production in time and space, and, its relationships with the environmental, educational and environmental education areas. This information permitted the identification of several topics and possibilities for study even when the matter was cross-curricular, the close relation with other knowledge areas like the physical sciences and the social sciences, besides emerging theoretical fields associated with particular fields such as health and indigenous matters.

The study allowed its authors to conclude that (2009: 230–231),

> The academic environment for investigation and discussion is more concentrated in the southern and southeastern zones of the country and this can be explained in two different ways: the existence of particular PhD programmes and the researchers' difficulties in accessing the area they wish to study and the costs of participation.

There is a large distance between the conservative vision and the appearance of a critical vision in environmental education with regard to those political and cultural aspects that constitute parts of a new epistemological position:

(a) The thematic and theoretical plurality brings together the environmental field and education, and the diversity of topics and emerging theoretical fields show the variations in complementary and inclusive knowledge.

(b) Some cross-curricular theoretical lines are present in the thematic plurality and they are the bases for discussing the concepts that link culture and nature. They include social representations, sustainability, subjectivity, phenomenology and hermeneutics.
(c) There is a defined interchange zone between the theoretical fields of ecology, culture, politics, citizenship and ethics.

With regard to ethics and education, the work of Paulo Freire is particularly relevant. For him, people must take responsibility over their own actions, what he calls "ética universal do ser humano". It is in the name of this ethics "...always together with the educational practice, does not matter if we were working with children, teenagers or adults, we shall fight" (1996: 17). Fighting is a way of participation for society dynamism, configuration of a sociability sustained by an educational process committed to generate more citizens involved in the protection of life, according to Jacobi (2003).

Social environmental education, in Paulo Freire's terms, is directed to the formation of a critical consciousness about what is experienced that promotes "the report of how we live and the disclosure of how we could live" (quoted in Pitano and Noal 2009: 293). Human action gives identity to people because it establishes their relationships with their equals and the world. People engage in self-directed, changeable actions. However, this is not a universalised, continuous and constant practice.

Actually, only a small part of society has the conditions to adopt activities that are in fact do-able. Because of the ignorance of their own possibilities or because of oppression, most of the people in the world are marginalised from the decision-making processes at all levels. It is a situation that must be changed, a task for the educational process, formal or informal, at school and at home, because through liberation education it will be possible to make changes in consciousness and thinking, modifying the submission and conditioning processes.

Education has a political dimension that not only leads to a new understanding about oneself but also for citizens, leading to new points of view and attitudes, opening their minds to a new way of thinking and acting in their everyday lives. We refer to Gadotti's convictions in *Pedagogia da Terra*,

> ...education would bring more and more power for the economic, political and social sustainability fight...the engagement and the ethics-political formation of larger parts of the public opinion are essential to make a more solid process flow and create better social conditions that allow economic and social sustainability (2000: 87).

For Jacobi (2003), students must be supported in order to position themselves in this complexity. The progress presented in this chapter via highly regarded authors writing about the same topic does not leave any doubts about the hybridism involved in environmental matters, such as development, diversification of problems, ideological implications, and economic and political interests.

Many dimensions of human life, individual or collective, are circumscribed by environmental dynamics in many more risky situations than anyone could have imagined. Environmental education, eco-development and sustainability are concerned with finding solutions to problems and these solutions cannot be found through the small and naïve initiatives of individuals or groups. Engaging in rational and conscious practices with regard to natural resources and adopting appropriate behaviour for rubbish disposal are very small contributions (although important and necessary) when we look at the whole environmental picture. They have little value if they remain discontinuous or isolated actions.

The educational process cannot avoid its responsibilities for contributing to the generation of good teachers especially in preparing them,

> to transform the information they get including the environmental aspect in order to be able to transmit and decode to the students the expression of the meanings about environment and ecology in its multiple orders and intersections. The emphasis must be on training to receive the relations between the branches as they were once, pointing out the local/global formation trying to emphasise the necessity of facing the logic of exclusion and differences. In this context the administration of the social and environmental risks points to the necessity to amplify the public movement by initiatives that would allow an increase in the level of environmental institutional consciousness of the inhabitants, assuring formation and consolidation of open channels to a participation in a pluralistic perspective…the teachers' role is essential to impel the changes to an education that makes a commitment to sustainability formation, as a part of a collective process (Jacobi 2003: 199 and 204).

Conclusion

The educational process has to be more efficient as a response to a competitive situation. State education must adopt the more effective techniques employed in the private sector. Through educational practices, people must be enabled to make their own choices. The rights to education, to citizenship, to having a high quality of life in a healthy environment, cannot be just rhetorical elements. In thinking seriously about citizenship, we must understand it as: civil rights practice (right to live, to freedom, to property, to equality before the law); and social rights that guarantee the inclusion of the individual in the collective wealth (education, work, fair salary, health, a peaceful old age) (Pinsky and Pinsky 2003: 9).

Referring to the teaching and learning process, Favaretto affirms that, "concerning the most important signs for education are the ones of human experience that imply culture and history…in the different cultures there are different dimensions of the temporality experience" (Lecture at the University of São Paulo – 4/6/2000).

The common knowledge is a product of people in a particular situation. At first, it is an intuitive understanding without thinking about reality and to improve on this there is an important educational function: to integrate culture, to explain events and to help learners to know and understand different social roles. That does not make it a neutral knowledge, with a lack of interest that is marginalised with regard to political and social questioning. The production of knowledge is present in everyone's daily lives, as we can observe in the researches and references used in this chapter. Humankind is historically open to receive influences from the time and the space in which they exist. Paraphrasing Japiassu (1991: 85), knowledge production "…takes place in a determined society that conditions its aims, its agents and its working process..." or, as Cortella (2000: 99) asserts, "...knowledge and on it the truth...are results of the effort of a determined group of men and women to construct references that guide the meaning of human action and existence".

Our discussion of the principal themes addressed in this chapter, drawing on the work of various authors and published official documents, affirms the importance of the following:

- understanding the topic as a product of effort of the created conditions and of the resources to be socialised and not as an isolated action of the more capable ones, of the more sympathetic individuals, or of the militants; such understanding must be reachable for all members of society, including the less active ones;
- with regard to the environment, there is a need for clearer thinking about the knowledge of environmental matters, such as nature (what it is), its sources (its origin) and its use (purpose);
- promoting educational relationships and the education-learning process, as a significant practice in a personal, political and social perspective, because the survival of the planet depends on that kind of consciousness in everyday experience that only educational institutions can promote.

Education linked to environmental matters should aim at developing competencies defined by development needs and, in specific subjects, should guarantee an understanding of basic concepts, policies and projects dimensions.

References

Acselrad, H (ed.). 2001. *Sentidos da Sustentabilidade Urbana*. Rio de Janeiro: DP&A.

Bauman, Z. 2003. *Comunidade: A Busca da Segurança no Mundo Atual*. Rio de Janeiro: Jorge Zahar Ed.

Camargo, A L de B. 2003. *Desenvolvimento Sustentável: Dimensões e Desafios*. Campinas: Papirus.

Comissão Mundial sobre Meio Ambiente e Desenvolvimento (CMMAD). 1998. *Nosso Futuro Comum*. Rio de Janeiro: Fundação Getúlio Vargas.

Cortella, M S. 2000. *A Escola e o Conhecimento: Fundamentos Epistemológicos e Políticos*. São Paulo: Cortez.

Foster, J. B. 2005. *A Ecologia de Marx: Materialismo e Natureza*. Trans. Maria Teresa Machado. Rio de Janeiro: Civilização Brasileira.

Freire, P. 1996. *Pedagogia da Autonomia*. São Paulo: Paz e Terra.

Gadotti, M. 2000. *Pedagogia da Terra*. São Paulo: Petrópolis.

Gaudiano, E G, and L Lorenzetti. 2009. Investigação em educação ambiental na América Latina: mapeando tendências. *Educação em Revista/Universidade Federal de Minas Gerais, Belo Horizonte: FAE/UFMG* 25(3): 191–233.

Gonçalves, M A R. 2003. *A Vila Olímpica é Verde-e-rosa*. Rio de Janeiro: Editora FGV.

Jacobi, P. 2003. Educação Ambiental: Cidadania e Sustentabilidade. *Cadernos de Pesquisa* 18: 189–205.

Japiassu, H. 1991. *As paixões das ciências*. São Paulo: Letras e Letras.

Jimenez, S, and E Terceiro. 2009. A crise ambiental e o papel da educação: um estudo fundado na ontologia marxiana. *Educação em Revista/Universidade Federal de Minas Gerais, Belo Horizonte: FAE/UFMG* 25(3): 299–332.

Pato, C, L M Sá, and V L Catalão. 2009. Mapeamento de tendências na produção acadêmica sobre educação ambiental. *Educação em Revista/Universidade Federal de Minas Gerais, Belo Horizonte: FAE/UFMG* 25(3): 213–233.
Pinsky, J, and C B Pinsky (eds.). 2003. *História da Cidadania*. São Paulo: Contexto.
Pitano, S de C, and R E Noal. 2009. Horizontes de Diálogo em Educação Ambiental. *Educação em Revista/ Universidade Federal de Minas Gerais, Belo Horizonte: FAE/UFMG* 25(3): 283–298.
Reigota, M. 1999a. *A Floresta e a Escola: Por uma Educação Ambiental Pós-moderna*. São Paulo: Cortez.
Reigota, M. 1999b. *Ecologia, Elites e Intelligentsia na América Latina: Um Estudo de suas Representações Sociais*. São Paulo: Annablume.
Sachs, I. 1986. *Ecodesenvolvimento: Crescer sem Destruir*. São Paulo: Vértice.
Santos, M. 2002. *A natureza do Espaço*. São Paulo: Edusp.
Scotto, G, I C de M Carvalho, and L B Guimarães. 2009. *Desenvolvimento Sustentável*. Petrópolis: Vozes.

Chapter 3
Schooling for Sustainable Development in Chile

Fabián Araya Palacios

Introduction

In the context of the social sciences curriculum for primary and secondary schools in Chile, this chapter analyses various aspects of schooling for sustainable development. In geographical education, there is an increasing interest in education for sustainable development and this is connected to the training of geography teachers where official publications and the use of new technologies are two valuable components of this new trend. The chapter is divided into three sections. The first section presents an overview of geographical education in the country. It outlines the current situation of geographical education with reference to the situation of geography teaching in higher education, teacher training and the situation of geography in the curriculum of elementary and secondary education. The second section deals with geographical education for sustainability and citizen formation. The third section presents some challenges faced by geographical education in Chile.

Since the 1980s, with regard to the broad field of the social sciences in the primary and secondary school curriculum, geography has played a secondary role, following behind history. However, educational reforms in Chile have transformed not only the curriculum system but also the teaching and learning activities for primary and secondary schools. Environmental issues and sustainable development have allowed geography and geography teachers to play an increasing role mainly because of the range of skills that these teachers bring to the study of geographical regions and the relationships between physical, economic, and human components of any landscape.

F.A. Palacios (✉)
University La Serena, Santiago, Chile
e-mail: faraya@userena.cl

M.L. de Amorim Soares and L. Petarnella (eds.), *Schooling for Sustainable Development in South America*, Schooling for Sustainable Development 2,
DOI 10.1007/978-94-007-1754-1_3, © Springer Science+Business Media B.V. 2011

Geographical Education for Sustainability: Background

The concern for geographical education in Chile dates back to the nineteenth century when the government in the Republican period decided to create suitable institutions for developing geographical education in schools and universities and for promoting studies to describe the whole country. Towards the end of the 1980s, geographical education was taught in a descriptive way and it started to change in the 1990s. This is due, among other factors, to the process of professionalisation of teachers and the changes experienced in the initial teacher training programme of geography in the universities of the nation.

Geography teacher training in Chile has been developed mainly in universities and professional institutes. Full-time geography teachers are not provided for the first and second grades in elementary education. By contrast, in secondary education a qualified teacher is responsible for history, geography and social sciences classes. This teaching activity is oriented towards the explanation and analysis of the temporal-space dimensions of societies. Geographical education is permanently and explicitly present in the new curriculum arrangement. As a consequence new generations of students will acquire geographic skills enabling them to understand the territory as a human construction characterised by constant change. Geographical education focuses on sustainability and citizen formation. These two extremely important trends are evident throughout primary and secondary education. The first trend, geographical education for sustainability, responds to a demand that has profound social, economic and environmental connotations including globalisation, climate change, human development and biodiversity and these are important concepts representing both ethical and solidarity dimensions currently taught to the new generations. Geographical education for citizen formation empowers students to be responsible citizens who can adapt themselves to local environments and develop good social relationships.

Schooling for Sustainability

In an effort to focus international attention on issues of sustainable development, the United Nations General Assembly declared 2005–2014 as the Decade of Education for Sustainable Development with the United Nations Educational, Scientific, and Cultural Organisation (UNESCO) being responsible for its organisation (UNESCO 2004). While the Declaration's full impact on geographical education is not yet known, there is evidence that geographical educators have the potential to offer contributions to the educational goals related to education for sustainable development.

The broad goals of education for sustainability, understanding the interdependence of life on Earth and the effects of personal and collective decisions, were developed in the 1990s under several terms, including education for sustainability, education for a sustainable future and education for sustainable development.

In the United States, education for sustainability has roots in environmental education, but now spans numerous disciplines. Many describe education for sustainability as a process rather than as a prescriptive measure for meeting the challenges of a 'globalised' world. Ethical motivation, environmental and active citizenship, critical thinking skills, and reflection are all desirable outcomes of education for sustainability (Ray 2007).

Sustainable development and sustainability are loosely defined terms open to multiple interpretations. The United Nations Brundtland Report defines sustainable development as "development that meets the needs of the present without compromising the ability of future generations to meet their own needs" (World Commission on Environment and Development on the United Nations 1987: 43). Many definitions of the two terms reflect a forward-looking attitude. Herremans (2002; referred to by Ray 2007) defines sustainability as a process capable of being maintained over the long term and maps sustainability at the intersection of economic, social, and environmental values. Actually the sixteenth session of the UN Commission on Sustainable Development which started on 5 May 2008 focused on the thematic issues of agriculture, rural development, land, drought, desertification and Africa (United Nations et al. 2008).

Geographical education for sustainability aims at answering to a demand with deep social, economic, and environmental connotations: globalisation, climatic change, human development, (biological, socio-economic, and cultural) diversity, and sustainable development (Stoltman 2004), implying not only the individualisation of interrelationships but also the positioning of an ethical and supportive dimension with forthcoming generations (Duran 2005). Geographical education, as an educational approach, presents diverse possibilities in order to contribute to sustainability, especially from the perspective of the interrelationships between human beings and the territory they inhabit.

The subject matter associated with sustainable development and environmental education has been a major concern among biology and natural sciences teachers in Chile. This situation is currently changing since geography teachers are also becoming involved with environmental education. Geography teachers educated at the university level are increasingly incorporating topics related to sustainable development into the formal curriculum.

The Chilean School System and the Formal Curriculum

The Chilean educational system has four sub-systems. The first sub-system, nursery school education (pre-school level), provides for children aged 1–5 years; the second sub-system, primary general education lasts 8 years (children aged 6–13 years); the third sub-system, secondary school education, lasts 4 years (children aged 14–17 years). Finally, the fourth sub-system corresponds to higher education, aimed at obtaining professional qualifications and/or academic degrees (Muñiz 2004).

All geography courses in primary and secondary education are integrated into the area of the social sciences through learning areas and sub-areas. During the first cycle of primary general education (first to fourth grade), geography curricular contents are integrated into the learning area titled "Comprehension of the Cultural, Social and Natural Environment". In this area, social, geographical, and historical topics are integrated. Among the main geography topics, we can find cartography, political-administrative organisation of the regions in Chile and their human and natural characteristics.

In secondary education, geography is integrated into the learning area titled "History and Social Sciences." Topics are dealt with from a local and regional perspective, moving towards a comprehension of the national, continental, and global geographical environments. The geographical topics most relevant for sustainable development are:

- The geographical region and its systemic analysis
- The natural, economic, social, and cultural aspects of the different regions of Chile
- The issue of pollution related to an inappropriate use of natural resources
- Poverty and inequality among the population
- Natural hazards
- The search for urban and rural environmental sustainability.

Geographical topics can be contextualised according to the spatial and cultural realities of the students. Thanks to this flexibility, the curricular contents become more appropriate and contribute in a better way to the materialisation of a spatial way of thinking in the students (Stoltenberg 2004).

Geography for Sustainability in the School Curriculum

The interest in defining a curriculum adjustment in geography began when the Chilean Society of Geographic Science made a national public statement in August 2007, raising a concern about the geographic situation. The Chilean Society of Geographic Science repeated later its concern to the Social Sciences team of the Curriculum and Evaluation Unit of the Ministry of Education (MINEDUC). The lack of explicitness about geographical contents in the curricular proposal was its main concern.

After asking for further details from the main people involved, the Ministry of Education of Chile is elaborating a curriculum adjustment in several areas and sub-areas of teaching with the purpose of modifying the current curriculum. This measure will bring about some positive changes in the field of geographical education, above all, in the introduction of new geographical contents and skills in the school curriculum. During 2008 and 2009, the Social Sciences team in the Ministry worked hard to incorporate geographic contents and skills into the different levels considered in the curriculum adjustment.

The curricular adjustment takes into account a number of important aspects. In relation to the need for a major presence of geography in school teaching, the adjustment stresses its incorporation into all levels of the education system, from the first grade of elementary school to the fourth grade of secondary school (12 years altogether). The presence of geography comes into existence through the denomination of the learning area History, Geography and Social Sciences with the main objectives and compulsory minimal contents. This adjustment also fulfills the need for a clear sequence in the learning of geography, providing a spatial sequence and a progression in the geographic skills from first to sixth year of elementary school and then from seventh to fourth year of secondary school.

With respect to the weaknesses of geography in the curriculum and the lack of significant geographic themes in school training, we have to point out that the adjustment implements an up-to-date view of geography as a social science. In that adjustment, geography includes important themes for the students at each level of education. These themes include the spatial pattern of human occupation of territory, the interrelationships between society and nature, an holistic conception of the planet as the home of human beings, the spatial impacts of the globalisation process, global warming, migratory flows, fast urbanisation, the location of transnational companies, relationships between Chile and its region and the global economy, foreign treaties and Chile's impact on geographic space. This up-to-date view of geography is linked to historical, economic, and political aspects learned at every school level. It is crucial to mention that the emphasis on geographical themes is organised according to the concept of sustainable development and the training of citizens in geographical knowledge.

Current geographical education shows a great potential for the development and the consolidation of the principles of sustainability in conformity with the guidance of the Lucerne Declaration (International Geographical Union, Commission on Geographical Education 2007). The space perspective, that characterises geography as a social science, unites the analysis of the typical aspects of sustainable development within specific geographic spaces.

There is a very close relationship between geographical education and citizen formation that provides an enriching experience. The students' familiarisation with the territorial structure of the country and its link with relevant democratic institutions corresponds to an important school experience in their citizen formation. Maybe, this is because school represents one of the most widespread experiences, where communities have the possibility to bring some influence, more than other agencies, on contemporary human life experiences. At school, friendship is experienced and, at the same time, some essential virtues of life in society are acquired.

With regard to geographic abilities which are important for students' spatial development, the curricular changes have incorporated the following criteria:

> First, for the concept of spatial location, it proposes that teachers recognise a progression of geographic abilities from the first to fifth grade of the elementary level. Conceptually, it refers to the knowledge of spatial location and the distribution of natural, social, economic, political and cultural processes on the Earth's surface. At first these abilities have a general

geographical view of the Earth but then they progress to become a more detailed knowledge of the planet, helping students to identify different geographic regions.

Secondly, it proposes that students should become aware of how human beings transform geographical space and, in turn, it influences human life. This ability is highly developed between the seventh grade of the elementary level and the fourth grade of the secondary level. Through this ability the static and immutable way that geographic space is seen can be transformed into a dynamic vision that contributes to an understanding of its richness, diversity and complexity.

Thirdly, the curricular adjustment considers the systematic comprehension of geographical space. It refers to the capacity of explaining the spatial dynamics of a specific territory consisting of different geographical dimensions. This ability is highly developed between the seventh grade of elementary level and the fourth grade of secondary level. It starts from the establishment of simple relationships between elements moving on to a systematic vision that incorporates different variables and allows for the linking of their location to other planetary places.

In brief, geographical education has been constantly and explicitly incorporated into the curricular adjustment. It is important that new generations of students develop geographical abilities so that they will be able to analyse and understand a territory as a human construction. In the curricular adjustment, geography plays an important role in citizen formation in the twenty-first century.

Progress Map for Schooling for Sustainability

This section discusses the concept of progress maps in geographical education for sustainability in Chile. These maps describe the typical sequence of learning development in certain areas or domains considered essential components for the education of the students at each curriculum level. This description is simple and concise so that everyone can share this vision about how learning progresses in the course of 12 years of schooling. The purpose is to enable teachers, students and parents to understand the meaning of progress in a specific domain of learning. The maps establish the connection between curriculum and evaluation, pointing towards what is important to evaluate and giving common criteria to observe and describe qualitatively the learning achieved. They are not a new curriculum, since they do not promote other learning. On the contrary, it is hoped that the maps deepen curriculum implementation, stimulating the observation of key competences that have to be developed (Ministry of Education 2009).

The progress maps express learning expectation for a curriculum subject in the form of a continuum through primary and secondary schools. The desired competences are expressed in terms of seven levels of progress that make up the progress map, including examples of how students can demonstrate these competences with cartographic materials. Note is taken of the complexity of the construction process and how competences are agreed for each level, including mention of the difficulties involved.

The curriculum for History, Geography and Social Sciences has the purpose of developing students' knowledge, skills and attitudes, allowing them to organise an understanding of society, either through its history or in the present, and prepares them for responsible social behaviour. It is expected that students will be able to comprehend the connections between society and the natural environment and appreciate the significance of the environmental balance (Eflin and Ferguson 2001).

The curriculum also promotes the development of the capacity for identifying, researching and analysing with precision problems concerning historical, geographical and social reality. The learning of History, Geography and Social Sciences has been divided into three Progress Maps: Society in Historical Perspective, Geographic Space, and Democracy and Development. The first two maps explain the learning progression mainly related to the disciplines of History and Geography. The third map, Democracy and Development, describes the learning relating to political coexistence and the dynamics of sustainable development including the skills necessary for active citizenship.

The Spatial Geographic Progress Map

This map describes learning that progresses around three aspects developing in an interrelated way:

- ***Spatial location and systematic comprehension of geographic space***: this refers to the knowledge of the location and spatial distribution of elements and geographic processes and the comprehension of spatial dynamics of a certain territory, incorporating several variables such as natural, social, economic, political and cultural variables. This aspect starts from a general geographic vision of the Earth and the identification of simple connections between geographic elements; and then it evolves toward a more detailed and systematic knowledge of the planet and understanding of the interrelations between different variables in the shape of geographic space (Lidstone and Williams 2006).
- ***Analysis abilities of geographic space***: this refers to the development of abilities related to the direct observation and interpretation of geographic space or through different information sources, applying geographical categories of increasing complexity, in order to analyse significant geographical problems, formulating hypotheses about their causes and their territorial impact (Batllori 2002).
- ***A valuing and responsible attitude toward geographic space***: this refers to the development of attitudes of being careful with and responsible for the geographic space, realising its role in environmental sustainability and in valuing belongingness to places, not only one's town, but also the whole planet. In the map "Geographic Space" there is the assumption that territorial order is a human construction, which can be modified for the benefit of the quality of life (Stoltman 2004).

The maps illustrate learning in seven levels, from 1st grade of elementary school to 4th grade of secondary school. Each level is associated with what students are expected to be able to achieve at the end of certain school years. For instance, level 1 corresponds to the achievement expected from most of the children at the end of 2nd grade of elementary school; level 2 corresponds to the end of 4th grade of elementary school and so on every 2 years. The last level 7, describes the learning achieved by a student who stands out from others at the moment of leaving school, that is to say, a student who has exceeded the 6th level expectations. However, reality demonstrates that different level students coexist in the same grade. Because of this, the work is intended to determine the level of learning for each student and which aim they have to move towards achieving and then to guide the pedagogical activities for improvement.

The following section describes the Map of Geographic Space. At first, there is a synthesis of all the levels. Afterwards, each level is detailed, starting from its description, some examples of performance to show how this level of learning can be recognised, followed by one or two examples of worksheets given to students from several schools together with comments on why a student's worksheet is 'in' the level.

Levels for the Spatial Geographic Progress Map

Level 7. At this level, students are expected to be able to establish connections between several geographical variables in order to explain the spatial dynamics in a particular territory. They should be able to interpret and incorporate information from different sources and scales for analysing social phenomena and problems, taking into account social, historical and economic variables, as well as formulating hypotheses about causes and consequences. Students can recognise the relevance of territorial planning as a rationing instrument for taking up space and evaluate the scope of environmental policy.

Level 6. At this level, students can depict the location of a country and its region in the world and the historical changes in the shape of Chilean geographic space. They should study the interconnections between economic processes, the composition of geographical space and the dynamics of the population. Students are able to integrate and incorporate information from various sources and at different scales in order to analyse phenomena and spatial problems, taking into account social, historical and economic variables. They can set out the challenges for sustainability because of globalisation and assess the environmental policy and citizens' participation in these subjects.

Level 5. At this level, students are expected to be able to describe the world population with regard to its distribution and cultural diversity; as well, to be able to describe different regions in accordance with development indicators and demographic dynamics. They can acknowledge that territories take shape through processes of cooperation and conflict among societies and that the spatial dynamics of a territory are formed throughout history. Students are able to interpret information taken from different sources in order to analyse changes and trends in the configuration of geographic space. They can characterise social and environmental problems of big cities and value measures taken for improving the quality of life.

Level 4. At this level, the student distinguishes the spatial distribution of natural processes on the Earth and recognises that Earth's existence is related to the dynamics of the geosystem. The student is able to understand that the adaptation processes of human beings have been altered over time, causing changes in the geographic space. The student can select sources of relevant information and interpret geographical information to analyse the dynamism, magnitude and spatial scope of natural and historical processes. The student is able both to comprehend that the Industrial Revolution produced a great impact of human activity on the planet and to value the current environmental conscience.

Level 3. At this level, the student is able to recognise the outstanding natural and human features of America and Chile. The student is able to understand that geographic space is distinguished by establishing connections between natural and human features. The student can extract information from regular and thematic maps and from written and visual sources in order to determine the natural and human features of certain geographic spaces. The student is able to recognise the importance of reducing the negative consequences of human activity on the environment.

Level 2. At this level, the students can locate Chile and its neighbouring countries on the political map and distinguish the major natural zones of the country. They are expected to be able to recognise the existence of different types of human settlements and possible complementary relationships between them. They can use images and simple texts for describing the characteristics of different human settlements and their connection with the environment. They are able to understand that those human groups always affect the places where they live.

Level 1. At this level, students can locate continents, oceans and climatic zones of the planet. They are able to use relative location categories and cardinal points to get bearings in geographic space. They can utilise images for describing observable features of different landscapes, making some simple connections of proximity, direction and distribution.

Examples of, and Comments on, the Levels

Level 1. How is it possible to recognise this level of learning? When students have achieved this level, they can carry out activities such as the following:

- Notice the climatic zones in the globe.
- Identify continents and oceans on the map.
- Follow a simple route with a map of their neighbourhood.
- Describe displacement in space using familiar points of reference and categories of relative position (e.g. right, left, up and down).

Examples of Students' Worksheets

The task: The students received a city map showing different buildings, as well as public and private places (e.g. a person's home). The map included a compass card. Afterwards they had to answer four questions. In two questions they had to locate different elements in the map using cardinal points and relative points of reference. The third question described a route, and students had to explain where the route led to. Finally, in the fourth question, students had to give instructions to move from one place to another.

Example of a Worksheet at Level 1

Look at the map and answer the following:

- If Juan leaves his home, he goes up the road until he gets to "Las Camelias" Street, he turns right there and then he continues walking. At whose house does he arrive?
  Answer: "Felipe's house".
- Felipe will visit his newborn brother at the hospital. Write the instructions in order for Felipe to arrive at that place.
  Answer: "I walk two blocks west, I walk two blocks south".

Comment: By tracing out the route to be followed and identifying Felipe's house, the student demonstrates that he uses relative points of reference to be located on the map. Moreover, by giving instructions using cardinal points west and south, he also shows that he is able to find his way and to trace out routes on the map, giving instructions that include the cardinal points.

Level 2. At this level, the students can locate Chile and its neighbouring countries on the political map and distinguish the major natural zones of the country. How is it possible to recognise this level of learning? When students have achieved this level, they can carry out activities such as the following:

- Point out where Chile is on a map of America, and colour the neighbouring countries.
- Compare general features of human settlements through images and texts (location, size, types of building, productive activities, and routes).

Level 3. At this level, the student can recognise outstanding natural and human features of America and Chile. The student is able to understand that geographical space is distinguished by establishing connections between natural and human features. The student can extract information from regular and thematic maps and from written and visual sources in order to determine the natural and human features of certain geographical spaces. The student is able to recognise the importance of reducing the negative consequences of human activity on the environment. How is it possible to recognise this level of learning? When students have achieved this level, they can carry out such activities as the following:

- Describe the connections between the productive activities of a place and the possibilities provided by the surrounding geographical environment.
- Use a physical map to describe the macro shapes of American relief.
- Locate the main concentrations of population on a map of Chile.
- Suggest solutions to problems arising from some of the negative consequences of human activities in their own region.

Example of a Students' Worksheet

The Task: Students received two maps of Paihuano, IV Region. One of them showed some human features of this town and the other showed some elements of the physical geography of the place (climate and rivers). It was possible to see clearly the relief of the town on both maps. Afterwards students received a text describing the main economic activities. They were shown two photographs of the zone depicting a mountain chain and a grape plantation for Pisco industry. Then, they were asked to mention the main physical and human features of the town, establishing connections between the landscape and the main economic activities developed there.

Example of a Worksheet at the Level

- According to what can be observed, what are the main natural and human characteristics of the geographic space presented?
  Answer: "Human characteristics: School of Monte Grande, School of Paihuano, populated centres and industries: Tres Erres, Pisco Capel, Pisco Control and Los Nichos. Natural characteristics: bottom of a valley plain, winding river of shallow and still waters, slopes of mountains with vegetation, areas of tablelands …"
- Explain the main relationships that may exist between the characteristics of this landscape and the economic activities developed.
  Answer: "The relationship that may exist is the fact that from the grapes which are produced there humans transform them into wine. Mountains have cracks at the surface so that, when it rains, water flows through them and naturally irrigates the grapes in those vineyards. Because of climate and the large amounts of agricultural activity there are plenty of vineyards".

Comment: The student is able to incorporate information from: maps (school and population centres), text (referring to industries), pictures (describing the landscape). She/he distinguishes the human and natural characteristics and interrelates them linking elements of landscape (such as climate and relief) with agricultural activity (the grapes which are produced there, humans transform them into wine).

Level 4. At this level, the student can distinguish the spatial distribution of natural processes on the Earth and recognise that its existence is related to the dynamics of the geosystem. How is it possible to recognise this level of learning? When students have achieved this level, they carry out activities such as the following:

- Use maps to describe the spatial scope of certain historical processes (for example, areas of influence of classical cultures, and the European expansion).

Example of a Students' Worksheet

The Task: Students wrote a short text about the eruption of Chaitén volcano in May of 2008 describing the evacuation of the city of Chaitén and making references to the extension of the affected region. In addition, they analysed a map that showed the superficial distribution of the Earth's tectonic plates. Then, they were asked to link the location of Chile with the existence of this geographical phenomenon and to determine others places where a similar phenomenon could occur.

Example of Worksheet at the Level

Look at the map (showing the distribution of the Earth's tectonic plates) and answer the following questions:

- What relationships can you establish between the location of Chile and the existence of volcano eruptions?
  Answer: "That Chile is in the limit between the Nazca and South African plates, which exert pressure of one plate on top to another, so it provokes huge energy and telluric movements".
- In which other regions of the planet is it possible to have a volcanic eruption? Why?
  Answer: "In areas such as Japan, on the islands near Australia and in areas of Asia near India, because in these locations there exist limitations between the plates that exert pressure and energy one plate on top to another".

Comment: The student, by applying his/her knowledge of the process of subduction and its link with vulcanism, is able to interpret the information given on the map. Thus, the student distinguishes the convergence areas from the divergence plates and locates areas of possible volcanic activity. So the student locates Chile in a convergence area of tectonic plates, referring to the Nazca and South African Plates and mentioning Japan, the islands near Australia and areas near India as areas of subduction.

Level 5. Students are able to interpret information taken from different sources in order to analyse changes and trends in the configuration of geographic spaces. They can characterise social and environmental problems of big cities and evaluate measures taken to improve the quality of life. How is it possible to recognise this level of learning? When students have achieved this level, they carry out activities such as identifying on maps the main cultural regions of the world.

Level 6. At this level, students can locate their region and their country in the World and the historical changes in the shape of the geographic space in Chile. They should understand the interconnections between the economic processes, the composition of the geographic space and the dynamics of the population. Students are able to integrate and incorporate information from various sources and at different scales in order to analyse phenomena and spatial problems, taking into account social, historical and economic variables. They can set out the challenges for sustainability because of globalisation and assess the environmental policy and citizens' participation in these subjects.

How is it possible to recognise this level of learning? When students have achieved this level, they carry out such activities as the following:

- Locate in a cartographical manner the main business flows for Chile at the international level.
- Interpret maps showing the transport and communication network to identify comparative advantages of some regions and places related to international business flows.

Level 7. At this level, students are expected to be able to establish connections between several geographic variables in order to explain the spatial dynamics in a particular territory. They should be able to interpret and incorporate information from different sources and scales for analysing social phenomena and problems, taking into account social, historical and economic variables, as well as formulating hypotheses about their causes and consequences. Students can recognise the relevance of territorial planning as a rationing instrument for taking up space and evaluate the scope of the environmental policy.

How is it possible to recognise this level of learning? When students have achieved this level, they can carry out such activities as the following:

- Explain how territorial planning can contribute to mitigating the damage to the population of natural disasters.

Results

The sequencing of cartographic materials in the teaching-learning process is always a relative question. The development of determined techniques makes progress possible in other more complicated areas thus forming a coherent chain. Of course, if we do not begin with a cartographic basis it is difficult to make progress in respect to complexity. In the progress map it is proposed to sequence ideas in respect of seven levels in the learning of geography.

Cartographic materials are one of the best resources geography teachers have for fostering conceptualisation processes in the classroom. Observing, decoding and interpreting the information on maps offers interesting teaching opportunities that closely reflect the job of progress maps. Graphic schematisation is a process of high didactic value, due to its quality of being a language simplifier in the processing of spatial information. Each one of the levels within a range of a competence is treated in its distinct possible projections starting from an ideal point of initiation in primary school and its possible development in secondary school.

In the map, integrated comprehension of several components of geographic spaces, more than the identification of isolated elements, is valued. There is also an assumption that geography comes into existence through the observation and interpretation of geographical processes in a particular territory, applying this principle to the learning of geography. Besides, the described progression of learning reveals that students recognise they live in a geographic space that is dynamic, interdependent and socially constructed, in which every person is responsible for the environmental sustainability of the planet and for the care of the places to which they belong.

The Education of Geography Teachers for Sustainable Development

The formal education of teachers is undertaken at both the university and professional institute levels in Chile. In the first cycle of primary education, there are no teachers committed to instruction in geography. The professional who teaches in the area of "Comprehension of the Cultural, Social and Natural Environment" is a generalist teacher with a training background involving all the learning areas (mathematics, language and communication, and arts, among others).

During the second cycle of primary education, teachers also have a general educational background. In some instances, they play a double role as a history and geography instructor. By means of an authorisation issued by the Ministry of Education (MINEDUC), teachers are the main formal instructors for primary education. The Ministry of Education is promoting a teaching programme for these teachers with a major in social sciences in order to allow them to deal in depth with the contents of history, geography, and the social sciences in general.

In secondary education, teachers give lessons in history, geography, and social sciences, with an education oriented towards the comprehension of spatial and temporal dimensions of society. There are some recent experiences focused on preparing history and geography teachers separately. However, the tradition of integrated history and geography is widespread in Chilean universities.

Teachers receive training courses provided by the Centre of Experimentation and Pedagogical Researches (CPEIP) during their in-service professional preparation, which depends on the Ministry of Education of Chile, by providing courses through personal attendance and by correspondence (e-learning). There are also training programmes focused on curriculum development with the support of universities oriented specifically to the discipline and didactics of geography.

There are textbooks specifically oriented to primary and secondary students, which are provided by the state, at no cost for students of municipal schools and state-supported schools. In addition, there is also a variety of textbooks produced by publishing houses available commercially. These include geographical topics with many illustrations and a variety of didactical activities.

The use of new information and communication technologies (ICT) by geography teachers to develop their lessons has been steadily increasing. As part of the same trend, schools have developed computer laboratories for their teachers and students. However, the equipment is generally insufficient given the permanent demands of students who must be assisted by their teachers. Geography teachers receive support in the pedagogical use of computers through a national project of the Ministry of Education of Chile called "Proyecto Enlaces" (Ministry of Education 2008). Through this project, diverse curricular innovations have been developed in classrooms, and didactical resources have been generated in order to teach geography by means of CD ROMS and on-line publications through the Internet. A specific example can be found in www.odisea.ucv.cl (Contreras 2008).

Among some Chilean journals oriented to geographical disciplines, and containing a section devoted to geographical education, we can mention: *Geoespacios*, from the area of Geography Sciences at the University of La Serena (www.geografia-uls.cl) (Geography Sciences 2008); *Revista de Geografía del Norte Grande*, from Pontificia Universidad Católica de Chile (www.puc.cl); and *Anales de la Sociedad Chilena de Ciencias Geograficas* (http://www.sociedadchilenadecienciasgeograficas.cl) (Sociedad Chilena de Ciencias Geograficas, 2008). Among the educational journals can be mentioned *Revista Educación Ambiental* (www.conama.cl/certificacion) (National Commission for the Environment CONAMA 2008a, b) which is published as a joint effort by diverse state institutions.

Geographical education in Chile is an emerging subfield which is supported by meaningful projects connected to the area of education for sustainability. From 2002, and due to the need for creating a coherent system of multiple experiences on environmental education developed from the State and civil society, the main target to be reached is the implementation of the *Sistema Nacional de Certificación Ambiental de Establecimientos Educacionales* (National System for Environmental Certification of Educational Centres, SNCAE) (Baquedano 2003). According to the National Commission for the Environment (CONAMA 2008a, b), the institutions committed to the SNCAE initiative are:

- Ministry of Education of Chile (MINEDUC)
- Corporación Nacional Forestal (CONAF)
- Comisión Nacional del Medio Ambiente (National Commission for the Environment) (CONAMA)
- United Nations Organisation for Education, Science and Culture (UNESCO)
- Asociación Chilena de Municipalidades (ACHM)
- Consejo del Desarrollo Sustentable (Council for Sustainable Development: CDS)

The SNCAE constitutes a joint work platform among diverse institutions which, due to its range and permanence in time, is destined to become a specific practice in order to face the challenge of education for sustainability. Through this programme, complementary courses of action will be developed to strengthen environmental education, to care for and protect the environment, and to generate an associated network for local environmental management. For that reason, the exchange of knowledge, the experiences shared by geography teachers from different parts of the world, and the constitution of academic networks, seem to be the most effective strategies for a world where distances are increasingly recognised in a virtual way.

Line of Development Environmental Education Programme

This line of development corresponds to an academic initiative belonging to the Social Sciences Didactics and Methodology Programme of History and Geography Pedagogics in the Social Sciences Department of the University of La Serena. Its purpose is to make contributions from investigations to the improvement of the education-learning process of geography.

General Objective: to develop fundamental and applied research in the ambience of geographical education in order to generate new knowledge that helps to improve teaching and formative processes in the different formal educational system levels.

Specific Objectives: To develop investigative processes in the geographical education area using an approach that is multi-paradigmatic with an integration of qualitative and quantitative methodologies.

To apply the results of investigation to the optimisation of the processes of education-learning of geography in selected cultural contexts and according to the needs of students of undergraduate and postgraduate courses.

To disseminate the results of investigations through specialist journals, monographs, participation in scientific events and cultural diffusion at local, national and international levels.

Fundamental investigation lines: Didactics of Geography, Cognitive Processes and Development of Spatial Thought, Curriculum Development in Geography, Geographical Education for Sustainability, Geographical Education for Citizen Formation.

Applied investigation lines: Strategies, Methods, and Innovative Techniques in order to teach geography, New Information and Communication Technologies applied to geographical teaching-learning processes, Elaboration of Multi-didactic materials to teach geography.

Challenges of Geographical Education for Sustainability

The challenges analysed are the following:

Academic and Teacher Training in Geographical Education: teacher training in geography represents a critical responsibility. The need for an improvement in the training processes of both professors and teachers is clearly evident at the graduate level (e.g. Masters and PhD levels).

Specific Investigation Project: Construction of concepts, capacities and spatial skills for the formation of a geographically informed citizenship. This project is carried out in collaboration with the Improvement Centre, Experimentation and Pedagogic Investigations dependent on the Department of Education of Chile. Students have developed the seminar called "The city of La Serena as a didactic space for the construction of the concept of geographical location, in pupils of the fifth elementary grade".

Geographical education research: several research projects related to geographical education have been developed. It is worth mentioning research projects in science and technology, projects funded by the Pan American Institute of Geography and History (PAIGH), and projects about global geographical education, sponsored by the National Science Foundation (NSF–U.S.A.).

Application projects: participation in the curricular revision of programmes of History, Geography and Social Sciences for elementary and secondary education is a project that has been developed in the context of the updating of the national curriculum promoted by the Unit of Curriculum and Evaluation based in the Ministry of Education of Chile. It has been developed in a multidisciplinary way with a wide participation of teachers and specialists, considering the new trends in knowledge and the challenges that education in the twenty-first century must face (See: http://www.curriculum-mineduc.cl/ayuda/ajuste-curricular).

Making a map of progress on geographical space: this project consists of developing a map of progress of the skills and capacities that children and young people should achieve in the 12 years of schooling, from the first year of elementary school until the fourth year of secondary school (See: http://curriculum-mineduc.cl/curriculum/mapas-de-progreso/historia-y-cs-sociales/).

Publication of geographical education articles in specialised journals: geographical education is of great concern to Chilean faculties and teachers. As a result, professors make increasing efforts to publish and share resources at the national and international levels.

Production of teaching materials for geographical education: several teaching materials for geographical education have been produced for elementary and secondary education. Textbooks with full colour illustrations, maps, atlases, handbooks, CD documents and lessons as well as digital files are common resources utilised by college students.

The use of new technologies in geographical education: the use of new Information and Communication Technologies (ICT) represents a really important challenge for the dissemination of geographical knowledge to new generations of citizens. The use of new technologies to strengthen spatial skills associated with geographical education is a project realised in 2009 in collaboration with the Educative Computing Centre of the University of La Serena. The project consists of organising seminars with teachers of History, Geography and Social Sciences in order to incorporate diverse educational resources and new technologies such as Geographical Information Systems (SIG for the Spanish acronym) and Google Earth into geographical education (See: www.cieuserena.cl).

The development of networks collaborating on geographical education: new technologies have paved the way for the development of national and international networks in order to promote collaboration in a variety of geographical education fields. Teachers and researchers are steadily being accepted as members of the international community of scholars sharing common challenges and interests.

Participation in international research collaboration in the Americas: this is being undertaken in collaboration with the Association of American Geographers, and the Gilbert M. Grosvenor Centre for Geographic Education at Texas State University–San Marcos, with the purpose of developing studies and diffusion of information on various themes related to geographical education (See: http://aag.org/Education/aag/edu_project_main.cfm).

Latin American Geo forum about education, geography and society: this is a project that has been developed in collaboration with the University of Barcelona mainly through the Geocrítica Internacional and a number of Latin American Universities in order to exchange ideas, experiences and documents related to geographical education for sustainability (See: http://geoforo.blogspot.com/).

Academic links: geographical education for sustainable development has been developed over several years through a variety of academic links between institutions in many parts of the world. At national level, evident links are, for example, with the Department of Education, the Geographical Education Committee of the Chilean Geographical Sciences Society, the Master in Social Science Education of University of Bio-Bio and the National Association of Didactics of the History, Geography and Social Sciences.

At international level important links are those with: the Geographical Education Commission of the International Geographical Union; the Commission of Geography of the Pan-American Institute of Geography and History; the Department of Geography of the National University of Cuyo, Argentina; the National Pedagogic University of Colombia; the Master of Geography Education of the University of The Andes, Táchira, Venezuela; the Centre of Geo-didactics Investigations of Venezuela; Universidad Pedagógica Experimental Libertador of Venezuela; the University of Barcelona, Spain; National Council for Geographic Education; the Association of American Geographers; the University of Regina, Canada. The websites for these national and international links are: http://geoperspectivas.blogspot.com http://www.cartoeduca.cl/ and http://www.geopaideia.com/?q=node/2.

Conclusions

Geographical education offers a wide variety of possibilities that can help members of society learn about and understand sustainable development in Chile. Here, geography can play a profound role since it masters how to deal with important relationships between human beings and the territory they inhabit, both at the rural and urban levels.

This importance has been funneled by the environmental global concern which has reached Chile, as an advanced country open to globalisation. Two factors, among others, are clearly defined as the main causes for this change. First, there are environmental publications produced by governmental agencies that are sent to schools through official channels and scientific journals related to geographical education and the geographical disciplines. Secondly, information and communication technologies (ICT) are being applied in the school environments along with the development of the World Wide Web and the Internet.

We appreciate the contribution of teachers and faculties to strengthen geographical education in Chile. It is important to emphasise that such concepts as sustainability, citizen formation, and new information and communication technologies encourage stronger spatial recognition. This approach allows citizens to be integrated with a globalised world in a positive manner. All this action is accomplished by developing spatial abilities that are pursued to achieve a sustainable vision of natural and human resources.

References

Baquedano, M. 2003. La dimensión internacional en la educación ambiental chilena, *Revista Educación Ambiental.* 1(1): 13–15. http://www.conama.cl/educacionambiental/1142/article-34321.html. Accessed 3 July 2010.

Batllori, R. 2002. La escala de análisis: un tema central en didáctica de la geografía. *Iber* 32: 6–18. España. Barcelona.

Comisión Nacional del Medio Ambiente CONAMA. 2008. Revista Educación Ambiental. http://www.conama.cl/educacionambiental/1142/channel.html. Accessed 3 July 2010.

Contreras, D. ed. 2008. Unidades temáticas de Cuarto Año Medio. Odisea. http://www.odisea.ucv.cl/pags/fset/unidad4c.html. Accessed 3 July 2010.

Duran, D. 2005. El concepto de lugar en la enseñanza, Fundación Educa Ambiente. www.ecoportal.net/content/view/full/30984. Accessed: 3 July 2010.

Eflin, J, and D Ferguson. 2001. Environmental futures: educating for sustainability in the 21st Century. *Research in Geographic Education* 3(1&2): 3–31.

Geography Sciences at the University of La Serena. 2008. Revista Geoespacios www.geografia-uls.cl. Accessed 17 May 2009.

Herremans, I M. 2002. Developing awareness of the sustainability concept. *The Journal of Environmental Education* 34(1): 16–20.

Instituto de Geografía, Pontificia Universidad Católica de Chile. 2009. Revista de Geografía Norte Grande. http://www.geo.puc.cl/html/revista.html. Accessed 3 July 2010.

International Geographical Union, Commission on Geographical Education. 2007. Lucerne Declaration on Geographical Education for Sustainable Development. http://www.igu-cge.org/charters.htm. Accessed 3 July 2010.

Lidstone, J, and M Williams (eds.). 2006. *Geographical education in a changing world: past experience, current trends and future challenges*. Dordrecht: Springer.
Ministry of Education. 2008. Enlaces Proyect. http://www.enlaces.cl/index.php?t=44. Accessed 17 May 2009.
Ministry of Education. 2009. Ajuste Curricular y Mapas de Progreso para la Educación Chilena. www.mineduc.cl. Accessed 3 July 2010.
Muñiz, O. 2004. School geography in Chile. In *Geographical education: Expanding horizons in a shrinking world*, eds. A. Kent, E. Rawling, and A. Robinson, *SAGT Journal, Geocom 33, on the special occasion of the IGU 2004 Congress in Glasgow*. London: IGU/CGE with the Scottish Association of Geography Teachers.
National Commission for the Environment (CONAMA). 2008. Educación Ambiental para la sustentabilidad. http://www.conama.cl/educacionambiental/1142/article-28763.html. Accessed 18 May 2009.
National Commission for the Environment (CONAMA). 2008. Revista Educación Ambiental. http://www.conama.cl/educacionambiental/1142/channel.html. Accessed 19 May 2009.
Ray, W. 2007. Placing Sustainable Development in the Curriculum of a Cultural Geography Course in the United States, eds. S. Reinfried, Y. Schleicher, and A. Rempfler. Geographical Views On Education for Sustainable Development. Proceedings of the Lucerne-Symposium Lucerne: International Geographical Union-Commission on Geographical Education. Lucerne, Switzerland. July 29–31.
Sociedad Chilena de Ciencias Geográficas. 2008. Anales de la Sociedad Chilena de Ciencias Geográficas http://www.sociedadchilenadecienciasgeograficas.cl. Accessed 3 July 2010.
Stoltenberg, U. 2004. Sin información no podemos alcanzar un desarrollo sustentable. *Revista Educación Ambiental*. 1(2): 5–7. http://www.conama.cl/educacionambiental/1142/article-34321.html. Accessed 3 July 2010.
Stoltman, J. 2004. Scholarship and research in geographical and environmental education. In *Geographical education: Expanding horizons in a shrinking world,* eds. A. Kent, E. Rawling, and A. Robinson. SAGT Journal, *Geocom 33, on the special occasion of the IGU 2004 Congress in Glasgow*, 12–25, London: IGUCGE with the Scottish Association of Geography Teachers
UNESCO. 2004. Decade of Education for Sustainable Development. http://portal.unesco.org/education/en/ev.php-URL_ID=27234&URL_DO=DO_TOPIC&URL_SECTION=201.html. Accessed 3 July 2010.
United Nations, Department of Economic and Social Affairs, Division for Sustainable Development. 2008. http://www.un.org/esa/sustdev/index.html. Accessed 3 July 2010.
World Commission on Environment and Development. 1987. *Our common future*. New York: Oxford University Press.

Chapter 4
Education and Sustainability: Figurations and the Ethics of Care – Bolivian Experience

Ubiratan Silva Alves and Maria Beatriz Rocha Ferreira

Introduction

In recent decades there has been a much greater international concern with climate. The greenhouse effect and its influences on the environment are a constant concern of international organisations, such as the Food and Agriculture Organisation (FAO) which highlights global issues related to food, referring to the impact on food supplies of pollution of rivers and seas, forest clearing, soil erosion and unplanned production. Some important international conferences have been organised to discuss these and other environmental issues such as Eco–92 in Rio de Janeiro (1992), Kyoto (1997) and Copenhagen (2009). On the one hand there has been progress in some subjects that has contributed to a greater awareness of the problems of the planet, but, on the other hand, changes in the behaviour of individuals will take years as will substantial change in social, economic and political patterns.

In this chapter we address a number of themes that are interrelated with issues of education and sustainability, such as: processes of incorporation of knowledge, social figuration; global problems; concepts of sustainability; and the ethics of care. These themes contribute to a reflection on education in Bolivia, modes of commercial production and the processes of migration where people are seeking a better life.

We, who live on the 'planet', arrived at where we are for various anthropological, philosophical, historical and sociological reasons. There is now an urgent need for change; otherwise the forecast is for mass extinction, a dying planet. The Earth has always suffered storms, climate changes and earthquakes. The species have suffered

U.S. Alves (✉)
Faculty of Physical Education, University of Campinas (UNICAMP/SP), Campinas, São Paulo, Brazil
e-mail: ubiratan@usp.br

M.B.R. Ferreira
Faculty of Physical Education, University of Campinas (UNICAMP/SP), Campinas, São Paulo, Brazil

M.L. de Amorim Soares and L. Petarnella (eds.), *Schooling for Sustainable Development in South America*, Schooling for Sustainable Development 2,
DOI 10.1007/978-94-007-1754-1_4,

and have adapted to the continuation of life. The human species experienced unique structural and functional modifications that led to the development of intelligence and a capacity for learning and socialisation. The species has been able to build societies, cultures, civilisation and history. But with time humans became anthropocentric and distanced themselves from other species, unlike the ancient societies, the indigenous and/or aboriginal people who consider themselves integrated with other species and the cosmos.

Legislation, disclosures by the media and various programmes undertaken by the government and non-governmental organisations on environmental issues appear to be having some effect on education and the behaviour of people in some parts of the world. However the process of producing sustainable change is slow and lengthy. The sociologist Norbert Elias, in his figurational theory, suggests that social changes are processes that occur over the long-term (centuries), unplanned and unintended (Elias 1993, 1994). Changes in the structural organisation of society, in the structure of behaviour and in the psychic constitution occur slowly and are interrelated. The behaviours already embedded in society are called social *habitus*. The social norms are internalised more deeply by the people to operate not simply at the conscious level and as a choice but in those layers below the levels of rationality and conscious control. In his theory, Elias argued that there is increasing social pressure on people to have more permanent self-control over their feelings and behaviours.

The social pressure occurs through different mechanisms such as the establishment of laws, rules, restrictions, force and/or psychological pressures. For instance, the feeling of 'shame' is one of the mechanisms of social control developed in the process of acquiring new behaviours. People started feeling ashamed to spit in public, a behaviour that was acceptable without any social constraint in the Middle Ages and later. Nowadays many people feel ashamed to throw garbage anywhere, an act that was acceptable many years ago.

The processes of formal and informal education have a decisive role in making up the individual identity (I-identity) and a collective identity (We-identity). However, I-We identity cannot be seen separately, or rather the individual and society as if they represent two separate objects. In order to bridge this gap, Elias introduces the concept of social figuration, that is, "...people through their inclinations and basic provisions are geared to each other and joined to each other in many different ways". He continued, "These people form networks of interdependencies or figurations of many kinds, such as families, schools, cities, states or social strata" (Elias 1999: 15). Power is in each figuration, and he writes, "...power is really nothing other than a somewhat rigid and undifferentiated expression for the special extent of the individual scope for action associated with certain social positions, an expression for an especially large social opportunity to influence the self-regulation and the fate of other people" (Elias 1994: 50). We all have power in society, the nature and extent of the scope for decisions open to us depend on the structure and the historical constellation of the society in which we live and act. Power is changeable in the process, which moves forward and back, leaning first one way and then another. "This floating balance is a structural feature of the flow of each figuration" (Elias 1999: 143).

Creating a social and individual *habitus* on sustainability of a civilisation with a greater awareness of the problems of the planet could take generations. The concept of civilisation in the thought of Elias is a consequence of a long term process consisting of a network of functional interdependence (Elias 1994). This interdependence permeates people's actions, thoughts and emotions, and in this way it breaks through the dichotomies of individual and society, they are part of the same link and structure.

The individual incorporates behaviour from social pressures imposed in their society. In turn society is made up of individual figurations and the relations of dependence among them. Several social norms have arisen in the interest of specific groups, but, Elias (1993, 1994) states, none of these groups or persons has control over the direction and consequences of social events. This brings us back to the notion of interdependence where the universe of actions taken, directions and results cannot be predicted by any of those involved. The planetary problems and solutions for education and sustainability should be looked for in the forms of interdependence. The challenge of Elias is for a 'society of individuals'. The human being needs to leave his or her anthropocentric place in society for a more integrated action with other species, the planet and the cosmos.

Concepts of Sustainability

A concept is an abstract idea of a thing, a being or a phenomenon and it is actualised under new information. Current concepts of sustainability and education are being updated through the influence of different factors, such as aspects of geology, weather, scientific knowledge, socio-economic changes, expansion of human consciousness and ethics.

Authors such as Boff (1999), Gadotti (2005) and Benfica (2008) present different concepts of sustainability and human development and make proposals for an eco-pedagogy, and their main ideas have been used in this chapter. The cultural, ethical, political, social, environmental and economic dimensions of sustainability were incorporated in the concept of sustainable development and this was used for the first time in the United Nations General Assembly in 1979 (Benfica 2008). Benfica states that this concept is used worldwide, disseminated by the Worldwatch Institute reports in the 1980s and especially by the publication of the report *Our Common Future* by the United Nations Commission on Environment and Development in 1987. This concept has received numerous criticisms for its reductionism and its trivialisation.

Another term analysed by Benfica (2008) is 'human development'. According to him, this term has the advantage of placing human beings at the centre of development whose central axes are 'equity' and 'participation', and it opposes the neoliberal conception of development. In this context, one can add the terms 'sustainable human development' (Coraggio 1996: 10) and 'productive transformation with equity'. According to Benfica's analysis the concepts of sustainable development and human development are broad and denote vagueness. And, yet, the United Nations has in recent years started using the phrase 'human development' as an indicator of quality of life based on indices of health, longevity, psychological maturity, education, clean environment, community spirit and creative leisure.

Another issue is the idea that a 'sustainable society' should be able to meet the needs of today's generations without compromising the ability and opportunities of future generations. Gadotti (2005: 18) has criticised this concept, pointing out that, "The concept of development is not a neutral concept. It has a very specific context within an ideology of progress, which implies a conception of history, of economy, society and the human being". Gadotti assumes that sustainable development has no place in a capitalist mode of production, which always seeks profit. He proposes an economy ruled by compassion and not by profit. Compassion must be understood here in its original etymological concept of 'sharing the pain'.

The neoliberal system produces social inequalities, increases the number of excluded people, promotes a privileged minority, and neglects the ailments that destroy many aspects of social wellbeing in favour of individualism and economic perspectives (Gentili 1998). Another important criticism is that the concepts of sustainability should not deal separately with the social aspects of environmental issues. These are intertwined and need to be resolved in such a manner. Gadotti (2005: 17–18) argues for the interrelational aspects of society and environment in sustainability: "…it is not just to clean rivers, clean air, reforest the devastated areas to live in a better planet in the distant future. One should give simultaneously solutions to environmental and social issues".

The philosopher and theologian Leonardo Boff (1999) in his book entitled *To Know How to Take Care: Human Ethic, Compassion for the Earth* navigates in different cultures, drawing on mythological knowledge and practices of different civilisations, cultures and societies. The human essence and wisdom of the myths is more 'care' than reason and will. And caring involves love, compassion, feeling the pain of another and more than ever a differentiated education. The ethics of care should be at the core of human relationships, to bring life.

Human greed, poverty, social problems, global warming, hunger, abandonment of children, youth and adults, trash, pollution, food quality, public health, among other problems, reflect the 'carelessness and lack of care' and responsibility of individuals. The wisdom of caring inhered in human beings and thus in all cultures past and present could contribute to 'life on the planet'. In neglecting nature we neglect ourselves, making us sick and, ultimately, leading us to our own destruction (Boff 1999). *The Earth Letter* (1987),[1] in its first principle, asserts exactly that: "to respect, to care for the community of life."

[1]The Earth Charter is a document designed by the World Commission on Environment and Development of the United Nations in 1987. It gained momentum at the Earth Summit in Rio de Janeiro in 1992 with the objective of defending the interests of sustainable peace and socio-economic justice. It is a kind of code of ethics planetarium, similar to the Universal Declaration of Human Rights, which only focused on sustainability, peace and socio-economic justice. The document was completed in 2000 and was translated into 40 languages.

In one of his presentations, Boff[2] argued that the design of an educational model for sustainability involves two major dimensions which are mutually reinforced: first, a new approach focused on re-signification of nature, emphasising the Ethics of Care and; secondly, an understanding of the complexity of life processes and the interplay of human action on the balance of these processes. The transdisciplinary approach seems a better way to deal with complex systems of exponential behaviour.

Boff (1999) proposes that the first step in this model requires a reconstruction of the existing ethics around the idea of care, having as the starting point school children who should be placed in direct contact with the natural environment. This contact would contribute to the development of sensitivity with gentleness and an understanding of the hidden side of natural processes. According to Boff, the formation of these young people is as important as the re-education of adults who influence the family, the place to consolidate habits. These adults are still part of the active population in society where the biggest changes are yet to be accomplished with the utmost urgency. The learning of adults involves relearning processes and the deconstruction and reconstruction of mental models. The number of leaders should be expanded to lead the possible changes in existing economic models in society. There is a need to reverse the impending social and environmental collapse. Business and civil society organisations can be major players in this process, which has proved to increase with changes based on this worldview 'caregiver'. In this sense, to prioritise education for sustainability of human life on earth is to awaken consciences, to prepare leaders and citizens in general to enable the construction of a possible and essential new planetary utopia.

Bolivia: A Brief Educational Background

Bolivia has been recognised as a multiethnic and multilingual country since the Revolution in 1952. However the recognition process to be incorporated in different sectors of society takes a long time, and this is still going on. This situation drew the authorities' attention to all communities with respect to the cultural diversity of the country, but the lack of political continuity from one system to another has prevented the consolidation of coherent and sustained educational policies (Gumucio 1996). In presenting a brief review of education in Bolivia, we start with a consideration of the impact of the colonisation process. The ideas of the indigenous peoples were undervalued in educational projects, until more recently. They knew how to care for the environment; they perceived themselves as integrated with the ecosystem and the cosmos. They and their ways of life were misinterpreted and devalued for centuries.

The tradition in education in Bolivia, according to Vásquez (1991), dates back to around 1580 BC with the *Tiwanaku* whose architectural and artistic expressions were very advanced. These people spoke the *Aymara* language that persists today in

[2] http://www.fractalis.com.br/site/modules.php?name=Conteudo&pid=18

much of the Bolivian population. The education of this people is based on values such as solidarity and mutual cooperation.

The *Inca* Empire dominated the region by imposing on *Colla* and *Aymara* a social organisation of a collective nature that corresponds to what is called the modern socialist state. Education went through a training period where sages conveyed their knowledge to the nobility of the Empire through oral education, practice and experience in areas that were called *Yachaywasis*, the houses where ordinary people in these areas were not accepted. Then came a new kind of education aimed at the people encouraged by *Pachacutej* whose purpose was to impose the language of Cusco on everybody (Pottier 1983). The three main thrusts of Inca education still in place today in the lives of the indigenous peoples are: do not steal, do not lie; do not be lazy.

The boys who came to the capital Cuzco had 4 years of study and were instructed by *Rumasini* (oral teaching). The first year was to adapt and the second was dedicated to learning and religious liturgy. In the third year they were introduced to important knowledge of government and administration. In the last year they were trained for war and began to study history. The goal of education was to promote political consciousness in the individual for public actions (Vásquez 1991).

In the early colonial period, the harsh living conditions of the natives marked the destiny of their descendants. They were considered 'beasts' with no rights, a situation that started changing after 1516 when they were considered descendants of Adam and Eve by Pope Paul III (Lein 1991). According to Claure (1989), despite all the advances, education continued to be elitist. Spanish children and *crioullas* (*mestizos* – interbreeding of Spaniards and natives), but not the indigenous people, were allowed to be educated in convents and parishes. The Jesuits, nevertheless, continued to provide education for the natives, respecting their local culture. These first steps were quickly suppressed by the official policy of King Charles IV who in 1785 declared that he did not need philosophers but obedient and good servants. He authorised the establishment of schools to provide indoctrination and literacy once more for Spanish children and *crioullas*, but not for the indigenous people. However, the missions of the Jesuits continued to participate in the educational process of the natives, in order to contribute to solving local problems in the planting, hunting, and sustainability of the tribe and in encouraging art and crafts.

The second phase of education in Bolivia started after the expulsion of the Jesuits. Public education became important and immediate soon after national independence. Simón Bolívar (1825), the first President of the Republic, considered education to be the cornerstone of the new republic and the first duty of the government (Gumucio 1996). In 1827 an educational plan structured the primary and secondary levels and also the College of Arts and Sciences, the National Institute of Literature, Arts and Crafts during the period of Antonio José de Sucre Alcalá (1825–1828). However, the government of Andrés de Santa Cruz (1829–1839) prioritised the foundation of universities and technical courses leaving aside popular education (Gumucio 1996).

The National Convention of 1851 introduced for all citizens the right to free education under the supervision of the state not only at the literacy level, but also for developing the potential of each student. The intention was good but it failed to materialise. After an economic crisis, in 1859 the government introduced a policy to establish a common programme in all public and private schools, during the presidency of José María Linares Lizarazu (1957–1861). Education was divided into two levels: basic level and university level. However, the universities continued to maintain the privileges of the Spaniards and their descendants and this level of educational provision did not reach the lower classes and indigenous people.

The first formal education programme for teachers was created in Sucre in 1909 with important partnerships for future teachers (Vásquez 1991). The Institute for Teachers of Secondary Schools was founded in La Paz in 1917 improving significantly secondary education in the country. The following aims emerged for education from this new educational thinking:

1. to be utilitarian, pragmatic and practical;
2. to be scientific: to train students in observation, classification, description, experiment and induction, that is, the methods of natural science;
3. to enable the students to experiment, to make comparisons and to think for themselves;
4. to overcome the separation of boys and girls in different schools;
5. to be integrated and globalised based on the teaching method of students' interests;
6. to provide an aesthetic education in pleasant environments;
7. to be secular, without the influence of the church or any kind of religion and creed;
8. to make education the object of governmental policy.

Nevertheless, Bolivia's efforts to improve education did not benefit either farmers or indigenous people and this led to a 'pro-Indian national crusade' proclaimed by President Hernando Siles, when he took power on January 6, 1926 (Claure 1989). The teachers began to organise themselves into cooperatives and associations, leading to the creation of the National League of the Magisterium. This period was very rich with regard to legislation and curriculum reforms culminating in 1927 with the approval of the plan for creating a laboratory of Pedagogy and Experimental Studies focused on the psycho-educational opportunities for Bolivian children.

The city's schools acquired a good reputation but the rural ones had less priority. Until 1931, the schools only benefited whites and *mestizos* but little or nothing was offered to the Indians. Despite attempts to establish schools for Indians and introduce methods of teaching in their own language, they never managed to actually deploy these projects. Elizardo Perez discovered the creative power of the Indians and asserted in his thesis that the rural school should be for the Indian *Warisata*. He founded a school in 1931 in the heart of the *Aymara* community (Lein 1996).

The Rural Education Statute, promulgated in 1939, stated that schools must meet the needs and characteristics of each region, including all of the five sections: kindergartens, primary schools, secondary schools, vocational schools and special schools for the mentally handicapped (Claure 1989). Structural changes in the country occurred in the Revolution of 1952 with the nationalisation of mines, land reform, the introduction of universal suffrage and education reform and development.

Supported by President Víctor Paz Estenssoro during his first mandate (1952–1956), the main educational changes related to the formation of Bolivian citizens: encouraging their full potential according to their interests and in favour of the community; strengthening the biological values of the people by promoting a healthy lifestyle; helping to guide individuals with notions of ethics; and strengthening the feeling of nationalism while fighting regionalism (Juarez and Comboni 1994). These procedures met the needs of peasants and indigenous people leading them to improve their lives and also leading to the progress of the country.

In 1968, a new more radical reform was initiated called 'counter-reform', under the second mandate of President Rene Barrientos (1966–1969). This included four areas:

- School education cycles: 6-year primary school, 4 years of intermediate school and 2 years of secondary school with the baccalaureate degree as the terminal examination, and higher education (university level);
- Adult Education;
- Special Education;
- Informal classes and extension activities (related to society).

The president Alfredo Ovando Candia, during his third mandate (1969–1970) after the death of Rene Barrientos, reaffirmed the state's responsibility in relation to education seen as an inalienable right of the people. Another important moment occurred during the brief presidential mandate of Hernán Siles Zuazo (1982–1985) that attacked the problem of illiteracy and approved a National Plan for Literacy and Adult Education (Juarez and Comboni 1994).

Under the mandate of President Jaime Paz Zamora (1989–1993) the Ministry of Education presented a proposal for educational reform with an emphasis on teachers' salaries and teacher training; educational reform as gradual, progressive, consistent, global and participatory; education for work; women's participation; decentralisation of education services; and increasing the number of schools (Juarez and Comboni 1994). In order to implement this proposal, a contingency plan was drawn up with finance from the World Bank, UNICEF and UNESCO that included: rationalising and reducing night classes under national supervision; a system based on continuous training; and a wage policy that prevented disparity in salaries (Vásquez 1991).

The government of Gonzalo Sánchez de Lozada (1993–1997) came into power and approved an educational reform project that presented a new vision of the country continuing the previous government projects. Since 1995, the aims of education have included:

1. The formation of men and women in Bolivia, promoting the harmonious development of all in the interests of the community;
2. Strengthening national identity and celebrating cultural, historical values of the Bolivian nation in its multicultural richness and its many regions;
3. Encouraging skills drawing on the potential for art, science and technology, promoting the ability to deal creatively and effectively with the challenges of development at local, departmental and national levels;
4. Developing skills and competencies, based on the understanding of language and expression of thought through reading and writing, logical thinking through

mathematics based on learning as a progressive development of knowledge, science and technology associated with productive work that will contribute to improvements in the quality of life;
5. Building gender equality into educational arrangements by encouraging greater active participation of women in society;
6. Instilling in people the principles of political and economic sovereignty, territorial integrity and social justice, including the promotion of peaceful coexistence and international cooperation.

Currently, the Bolivian educational system is divided into four cycles, attending the childhood and youth development phases, as pre-school (4–5 year-olds), elementary education in the primary cycle (for 6–10 year-olds); intermediate education in the middle schools (for 11–13 year-olds); and 4 years of secondary education (for 14–17 year-olds) and university level. The 4 years of secondary education are divided into two cycles each lasting 2 years. The first cycle is a common core and the second one has some degree of specialisation in the humanities or in technical fields. The main goals of education are:

(a) training of the student;
(b) overcoming illiteracy; to promote justice, solidarity and social equality;
(c) promoting socio-cultural diversity and respect between the citizens through the support of intercultural activities and the recognition of the importance of different languages;
(d) promoting national integration and the participation of Bolivia in the world community based on sovereignty and national identity;
(e) contributing to the building of a democratic society in which everyone enjoys the same rights.

This very brief review of the educational history of Bolivia highlights the efforts the country has made in relation to organisation and laws. The situation continues to be critical not only for political reasons but also because of the lack of qualified staff to develop projects. The subject of education for sustainability still needs to be stressed in government policies. People with no experience in education occupy managerial positions in education, often hindering implementation procedures and local development. The National Department of Education has often exercised the power of routing issues in local areas (Barbosa Filho 2008).

The Bolivian Economy: Gas and *Coca*

The formation of the Bolivian ecological conscience has been strongly influenced by the colonisation processes, political and economic struggles, educational problems, the globalisation process, and so on. Unfortunately the local knowledge of the indigenous people apparently played little part in this process. Environmental issues, specifically gas and the *coca* leaf produced in Bolivian soil, are worth mentioning in this discussion (Hage 2007).

Agriculture accounts for approximately 11.3% of the GDP of Bolivia and the amount of arable land has been increasing, especially in Santa Cruz de la Sierra where the climate permits two crops a year. The extraction of minerals (zinc, silver, lead, gold and iron) and hydrocarbons (gas and oil) makes up another 14% of GDP. The national economy revolves mainly around the planting of *coca* and gas extraction, activities that directly affect the natural environment. Natural gas has emerged not only as the main source of revenue and export for Bolivia, but also as a source of political tensions involving regional governments, the central government, international companies and indigenous peoples (Hage 2007). This gas is found in nature in limited quantities and obviously diminishes with its exploitation. Once depleted, it can only be regenerated over a very long period of time. Moreover, the distribution of these resources on our planet does not occur uniformly, being concentrated in certain territories such as Bolivia. The country has the second largest natural gas reserves in South America (the first are in Venezuela) and it exports 50%. The President Evo Morales nationalised the control of the gas produced in the country, resulting in a reduction in investment by foreign enterprises (Barbosa Filho 2008).

The heavy rains in Brazil led to the increasing production of energy from hydroelectric plants resulting in a slowdown in the demand for gas from that country and this has impacted on Bolivia. The country can only increase its production of gas if Brazil and Argentina (currently its only consumers) continue to consume its gas (Hage 2007). One of the serious environmental impacts of extraction of natural gas in Bolivia can be observed in the *Guarani* community of Cumandaroti according to Hage (2007). The people there are directly affected by activities related to the gas and fumes, such as noise and water pollution and damage to agriculture and livestock.

The state of Tarija has a large gas production mainly in the province of Gran Chaco (one of six provinces in the state). The increase in extraction has caused conflicts between indigenous communities and the companies, especially with regard to the largest gas plant called Margarita. This region established an organisation to represent the Indians in these discussions, the *Itika Guasu*, representing these people in issues of environmental contamination of the Pilcomayo River and to defend the interests of the *Guarani* communities (Barbosa Filho 2008). The expansion of the gas extraction industry made the Bolivian government build roads that directly affect the wildlife habitat of the Chaco region.

Another supplier of informal work that affects the environment is the cultivation of the *coca* leaf, which employs thousands of workers in Bolivia. The *coca* leaves are used for teas, religious rituals, food and pharmaceutical products. However, there are negative consequences. Forests must be cleared for the plantations and unfortunately many countries use *coca* to manufacture the narcotic cocaine (Costa Neto 2005). According to Costa Neto, a planting of *coca* lasts 40–50 years and yields three-to-four harvests a year. *Coca* is the only crop that provides a lucrative business for the peasants of a depressed agricultural economy. Currently, at least 30,500 ha of *coca* are cultivated and, although it is difficult to assess the impact of *coca* cultivation on the Bolivian economy, it is clearly an important support to economic activities.

The cultivation of *coca* leaf and, more particularly, its processing into cocaine, has a negative impact on the environment. There is an increase in deforestation and soil erosion and in the dumping of toxic chemicals in streams and rivers due to the increasing use of pesticides to manufacture the drug (CEDIB 1986). A major barrier to reductions in the plantations lies in the fact that this agricultural activity may be considered as a cultural trait that meets the needs for economic survival of the people directly involved.

Coca cultivation in Bolivia has increased and consequently the production of cocaine as well. Bolivian anti-drug legislation stipulates that legal crops, for pharmaceutical uses, herbal tea, chewing leaves and religious rituals, are limited to 12,000 ha, at most. Meanwhile, President Evo Morales has emphasised that the limit should be 20,000 ha, remembering that for local use, or for use by Bolivians, 7,000 ha of plantation would suffice (Barbosa Filho 2008). Nationwide, 120,000 small farmers grow the plant, especially in regions of the Yungas (near La Paz), in La Paz and Chapare. Of the total national production, only 29% is used for purposes of local consumption with 71% destined for the manufacture of *coca* paste or powder cocaine. This production is sold to the *Chaques*, Bolivian middlemen who provide the plant. In Bolivia itself, part of the harvest is already processed for sale while another part is processed in Colombia along with the Argentines, Paraguayans and Peruvians who are responsible for distributing the goods (mainly to Brazil, that consumes 80% of Bolivian production) (CEDIB 1986).

The government emphasises that all *coca* excess should be discarded or subjected to processes of agricultural restructuring. The leaf that does not have a lawful purpose should be eradicated to avoid the manufacture of cocaine. Bolivia is the second largest world producer of *coca* leaves behind Peru. In relation to cocaine, it also ranks second, behind Colombia. Much of the cocaine is destined for the United States, stresses Domingues, Guimarães, Mota and Silva (2009). In Bolivia the use of the leaf has medicinal, cultural and religious purposes. It has been used as a stimulant and a preventitive against symptoms of altitude sickness, cold and hunger. There is a large market for this leaf in La Paz city, located in a very poor region. However, due to the audits of the government, getting the leaf is not easy. Consumers chew the leaf without the stalk and it has an effect similar to coffee consumption. They can also make tea, which they call mate. In addition to this destination, *coca* is used in the manufacture of liqueurs, sweets and wines. However, it is extensively used as a leaf chewed *chicle* to help the altitude adjustments, which facilitates the oxygenation of the blood when the thin air does not provide enough oxygen (CEDIB 1986).

The average price of *coca* leaves currently varies between US$2 and US$5 for half a kilo. The use of *coca* by Bolivians, mostly peasants, miners and workers, is to keep them awake and not hungry, so they can work longer and increase their incomes. Another use is for small children to prevent starvation, because there is not enough food for everyone in certain places.

Planting of *coca* in Bolivia has been increasing year by year and even the President Evo Morales is a *coca* farmer. He made his political career among the growers in the Chapare region, fighting against the eradication of crops and making roadblock protests that helped topple two presidents. In his speeches, he always claims to be in favour of *coca* but against cocaine (Barbosa Filho 2008). Chewing *coca* leaves was done 500 years ago. It is a habit of at least one million Bolivians, mostly peasants, Indians and prospectors.

Economic issues seem to dominate environmental and educational issues. However, there are some educational initiatives concerning issues related to the conscious use of natural resources, such as the FADEC (Foundation to Support Community Development[3]). It offers technical information to farmers both on economic and cultural issues and especially on issues related to social and sustainable development. Another organisation that works in the area of sustainability is the Pacha Gaia Foundation,[4] with the goals of care and preservation of the environment, through the promotion of initiatives and research projects, extension and development in different fields. In relation to drugs, there is an organisation called Bolivia Kotec[5] that works with drug users, offering legal and psychological assistance to users by promoting strategic actions in educational institutions.

The picture above highlights the need for consistent and integrated programmes with different sectors of society. One suggestion is the development of programmes based on 'ecopedagogy' (Gutiérrez and Prado 1998). This educational philosophy aims to build up an environmental citizenship based on caring for the environment and promoting learning about the meaning of everyday things. A similar approach is biopedagogy, 'Pedagogy of life' that offers more spiritual depth (Gadotti 2005). In this regard, Gadotti (2009) emphasises that ecopedagogy should grasp a much broader awareness, or planetary awareness, with a new ethical and social reference: a planetary civilisation. Eco-pedagogy, according to Gadotti conceives the human being in his or her diversity and in relation to the complexity of nature. The earth should also be regarded as a living being, like Gaia. Gadotti suggests using the term 'Pedagogy of the Earth' instead of 'eco-pedagogy'.

ADIÓS Bolivia

Many Bolivians leave the country looking for a better quality of life, partly because of land conflicts and the lack of employment opportunities. They leave Bolivia dreaming about a better life with better work and health. The main destinations of Bolivians today are the countries of South America, especially Brazil. Sidney Antonio Silva has engaged in deep and extensive research on the Bolivian immigants' way of life (Silva 1997, 1999, 2005, 2006). Some of the main ideas of this author are reviewed below.

[3] http://ong.tupatrocinio.com/fundacion-para-el-desarrollo-social-integral-de-bolivia-ong-2078.html

[4] http://ong.tupatrocinio.com/fundacion-gaia-pacha-ong-1954.html www.gaiapacha.org

[5] http://ong.tupatrocinio.com/kotec-ong-470.html

Migration to Brazil reached a landmark in the 1950s, when Bolivians went to study in Brazil through a Brazil-Bolivia exchange programme. However, from the 1970s, migration in Latin America has been related to the industrialisation of countries like Brazil, Argentina and Venezuela and to national, political problems. These movements are directly linked to a lack of public policies specially related to sustainability issues. The globalisation process has affected the economy of poor countries that have become more dependent on rich countries, failing to retain their own people on their own soil (Silva 1999).

From the 1980s there has been a large increase in Bolivians coming to São Paulo, no longer persecuted by authoritarian governments or for academic purposes. They have been people with low educational levels in search of work. This increase has been due to an economic crisis in Bolivia. This occurred with a process of reorientation of the workforce following the privatisation of the mining sector, which caused many layoffs. Adding to this, a rural exodus occurred affecting nearly 60% of the population who migrated mainly to La Paz, Cochabamba and Santa Cruz de la Sierra. These migratory movements are associated with the process of globalisation that has affected the economies of poor countries that were at the mercy of aid from rich countries and were failing to maintain their population on their own soil. The lack of public policies, especially those related to sustainability issues, affects these people (Silva 1999).

In general, the profile of Bolivians who travel to Brazil in search of work in recent decades, according to Silva (1997), is young, single, with an average level of schooling and mostly male. The female presence has increased considerably in recent years. Add to that list the independent professionals such as doctors, dentists and engineers, this is a significant contingent. All of those who have vocational skills, especially in technology, are welcome today. Silva (2006) asserts that immigrants seek better working conditions and some kind of income, which their own country is unable to offer and they dream of a better quality of life for themselves and their loved ones. In Bolivia, work opportunities are extremely scarce and offer little hope of promotion to those without a college education. The main goal is not to save money but to have better living conditions. About half of the Bolivian population is economically active in the informal economy and in addition to the problem of unemployment there are also problems of homelessness and saturation of public services, including health and education.

Silva emphasises that the most attractive work activity for the Bolivians is the business of sewing. This activity does not require much study or learning. What matters most in this type of activity is the ability to cope with the sewing machine and the possession of enormous physical strength together with the psychological strength to withstand the long hours of production. Most of the migrants work in sweatshops that seam the clothes of the city's textile sector. These small clothing manufacturers run by Bolivians are mostly family businesses which tend to grow and to use more advanced technologies. They are based on a network that recruits and hires predominantly undocumented and low paid labour (Silva 1999).

Silva stresses that this kind of life, immersed in sweatshops to achieve the goal of having an increased income, has serious consequences. Social, educational, cultural and leisure activities have virtually no space in the migrants' daily lives. In this activity there is a network for the hiring and recruitment of manpower where the established Bolivians stimulate their compatriots to migrate to Brazil. The employment relationship between the fashion designer and the owner of the workshop is very casual (the tailor normally receives payment by the produced piece), with no security or right in the laws governing the CLT (Consolidation of Labour Laws), making it vulnerable to the ups and downs of the market and the greed of the employers. Bolivians live and work normally on the same garments. They need to pay the boss for the sewing machine, housing, water, electricity and food. So, they end up in debt and are virtually 'stuck', which means that their employers lock the doors of factories and threaten to call the federal police to deport those who are illegal immigrants. This occurs despite Article 149 of the Brazilian Penal Code which considers it a crime for a person to be in a condition analogous to slavery, assigning a penalty of 2–8 years in jail and a fine for those responsible (Silva 1999). The situation is contrary to the words of Bolivia's National Anthem: "*¡morir antes que esclavos vivir!*" ("We will die before living as slaves!"). They came in search of a new lifestyle; however, many of them live almost as slaves in São Paulo.

The least the Brazilian authorities should do is facilitate the legalisation process. For legalisation of all documents, Bolivians have to pay an amount close to US$450, which becomes impractical for most. Other ways to legalise employment is marrying a Brazilian or having children born in that country. There are also 'amnesties' promoted by the government, which legalise the illegal immigrants. However, demand for such amnesties is not very large. There is a fear of contact with the police and the costs incurred together with the lack of information are considered to be explanations for the low number of beneficiaries.

In the last three amnesties granted by the Brazilian government the total numbers of all immigrants were 27,000 in 1981, 30,000 in 1988 and 39,000 in 1998 (Cymbalista and Xavier 2007: 128). Two years later, there was a shortage of at least 20% for registration confirmation. President Luiz Inacio Lula da Silva signed in 2009 the Law of Migration Amnesty which benefited about 41,000 migrants and the largest group was Bolivian, with more than 16,000 beneficiaries (Mendes 2010).

The living conditions of these groups are generally poor and degrading. The sweatshop is also the living space, i.e., the place to work, to eat, to sleep, to produce children and to live. The living conditions are extremely unhealthy. Alongside these bad living conditions the migrants find special inter-ethnic spaces for their leisure. They participate in events at the Memorial of Latin America in São Paulo, where they are able to meet with their compatriots. On weekends they can attend a cultural fair in the Plaza Kantuta, a square in the Canindé neighbourhood. The Plaza Kantuta receives 5000 people every Sunday, the majority being Bolivian, including natives and children who go to meet friends, to have fun, to enjoy typical food, to seek employment, to flirt and to share some of the customs from their country (Cymbalista and Xavier 2007: 127). In this place, the immigrants have created a 'piece' of Bolivia in Brazil to revive their culture and to re-signify their culture and values. In the centre of the square is a soccer field where they can play every weekend.

Games organised by clubs have become a key point in the lives of this group. According to the Pastoral Care of Migrants, there are more than 800 soccer teams from Bolivia in São Paulo, organised into 30 leagues.

Once every 4 months, in the Parish of Our Lady of Peace, a mass is organised with Bolivian songs and costumes. This church was declared the Faithful Latin Americans Parish in 1995, which is another spiritual and educational support for Bolivian migrants, helping them also in the amnesty process. It is located in the same area as the Centre for Migration Studies and the House of Migrants which provide the services of shelter, food, welfare and living space for those mostly from the African continent and the countries of South America. Beyond the church and the Plaza Kantuta, they organise sometimes other cultural events at the Memorial of Latin America.

In the field of health, records of the Basic Health Units of the central region of São Paulo have shown that the majority of patients affected by tuberculosis are Bolivians. The sewing work and the conditions of the workshops with little ventilation and a diet low in protein and vitamins are the main explanatory causes of these health problems. These *hermanos* are more likely to develop lung disease. Economic profit is gaining ground over education and sustainability and the ethic of care and compassion.

Final Considerations

The ethics of care, of compassion, responsibility and cooperation is a challenge for future generations. The actions are consequences of actors and institutions which are interconnected to form different figurations for education and sustainability. The formation of a social and individual *habitus* on sustainability in a society with a greater awareness of the problems of the planet will take a long time, possibly several generations. Making social changes is a long-term process and, according to Elias's theory, unplanned and unintended. Individual and society are not separate; they are part of the same link and structure.

The development of an awareness and attitude on education and sustainability is a consequence of public policies and private initiatives from different sectors of society. Political systems can be impediments to the development of planetary consciousness and attitudes. Many of them produce an increase in social inequalities, the number of excluded people, and the maintenance of a privileged minority in power, which destroys the various forms of social welfare, leaning towards individualism and a profit-driven economic perspective.

The Bolivian processes of economy, politics and education highlight the principles of a neoliberal policy. They reflect a class project that perpetuates the social exclusiveness of a privileged minority who for centuries has remained in power. The excesses of the Bolivian state are aided indirectly by the existing ethnic and cultural divisions in Bolivian society. The minority holders of wealth, who extract natural resources (mainly gas), and the small Bolivian farmers, who grow *coca* leaves that will target the production of cocaine, are far from thinking either about the environment in which they can consume natural resources or about the environmental needs of future generations. It seems that these groups would be hindered in their business if they had to consider the environmental impacts of their extractions and productions.

The immigration process is almost the last resort for the descendants of the native peoples of Bolivia, especially the *Quechua* and the *Aymara* as they seek to escape from the oppression and exclusion which has demeaned them for centuries. As they arrive in Brazil as illegal immigrants, this situation still worsens from the point of view of the ethic of caring for oneself, the other and for the planet. They stay often in a worse situation compared with their life in Bolivia. The Bolivians outside of their country are seen as 'Indians', poor in wealth and culture. According to Cavalcanti (2005), this image of Bolivians fits into a prejudiced concept of immigrants, building ideas that operate to marginalise groups. They occupy places in the lower social system and are blamed for all the ills that plague the regions in which they live. In this mindset, the Bolivians in São Paulo with their sounds coming from their Andean flutes can annoy people prejudiced against the sounds of drums, played in the slave quarters, that bothered the residents of 'Casa Grande' (a term referred to the House of Lords at the time of the slaves). This symbolic reality deserves to be rethought in the context of contemporary urban migrants and the recent migratory process.

A society where we recognise and respect multiculturalism, taking into consideration differences, overcoming racism and ethnocentrism still seems distant. The ethnic background of Latin America derived from indigenous peoples, Africans and Europeans must be revived to diminish the gap between them and to deconstruct the prejudices.

The philosophy of the *To Know How to Take Care: Human Ethic, Compassion for the Earth* or similar ethics are important fundamentals for programmes on education and sustainability. To care for others and for ourselves implies a multicultural dialogue which should be started in childhood. The incorporation of these elements by a society of individuals is a long-term process that may take generations, but it can be achieved.

References

Barbosa Filho, A. 2008. *A Bolívia de Evo Morales*. São Paulo: Editora Livro Ponto.

Benfica, G. 2008. *Sustentabilidade e Educação*. Seara (Salvador. On Line) Vol. 3:8.

Boff, L. 1999. *Saber Cuidar: Ética do Humano – Compaixão Pela Terra*. Petrópolis: Vozes.

Cavalcanti, L. 2005. Imigrante na cidade: paradoxos e pleonasmos. *Travessia – Revista do Migrante*, São Paulo: Centro de Estudos Migratórios, No 51

CEDIB. 1986. *Coca: Cronologia Bolívia*. Cochabamba: ILDIS – CEDIB.

Claure, K. 1989. *Las Escuelas Indigenales: Otras Formas de Resistencia Comunaria*. La Paz: Hisbol.

Coraggio, J L. 1996. *Desenvolvimento Humano e Educação*. São Paulo: Cortez.

Costa Neto, C. 2005. *Políticas Agrárias na Bolívia*. São Paulo: Expressão popular.

Cymbalista, R. and Xavier, I. R. 2007. A Comunidade Boliviana em São Paulo: Definindo Padrões de Territorialidade. *Caderno Metrópole* 17: 119–133, 10 set. 2007. http://web.observatoriodasmetropoles.net/download/cm_artigos/cm17_96.pdf. Accessed 31 Aug. 2010.
Domingues, J M, A S Guimarães, Á Mota, and F P da Silva. 2009. *A Bolívia no Espelho do Futuro*. Belo Horizonte: UFMG.
Elias, N. 1993. *O Processo Civilizador*, vol. 2. Rio de Janeiro: Zahar.
Elias, N. 1994. *A sociedade dos Indivíduos*. Rio de Janeiro: Zahar.
Elias, N. 1999. *Introdução à Sociologia*. Lisboa: Edições 70.
Gadotti, M. 2005. Pedagogia da Terra e cultura de sustentabilidade. *Revista Lusofóna de Educação*, No. 6, 115–29. Lisboa: Universidade Lusófonas de Humanidades e Tecnologias.
Gadotti, M. 2009. *Educar Para a Sustentabilidade*. São Paulo: Instituto Paulo Freire.
Gentili, P. 1998. *A Falsificação do Consenso*. Petrópolis: Vozes.
Gumucio, M B. 1996. *Breve História Contemporânea da Bolívia*. México: Fondo de Cultura.
Gutiérrez, F, and C Prado. 1998. *Ecopedagogia e Cidadania Planetária*. São Paulo: Cortez.
Hage, J A A. 2007. *Bolívia, Brasil e a Guerra do Gás*. Curitiba: Juruá Editora.
Juarez, J. M. and Comboni, S. 1994. Sistema Educativo Nacional de Bolivia: 1997 / Ministerio de Desarrollo Humano – Secretaría Nacional de Educación y Organización de Estados Iberoamericanos; *Revista Iberoamericana de Educación Número 5* Calidad de la Educación. May–August.
Lein, H S. 1991. *Bolívia: Do Período Pré Incaico a Independência*. São Paulo: Brasiliense.
Lein, H S. 1996. *Bolivia: The evolution of a multi-ethnic society*, 2nd ed. Oxford: Oxford USA TRADE.
Mendes. V. 2010. Brasil anistia 41.816 estrangeiros em situação irregular. *Agência Estado*. 8/01/2010. http://www.estadao.com.br/noticias/nacional,brasil-anistia-41816-estrangeiros-em-situacao-rregular,491657,0.htm. Accessed 31 August 2010.
Pottier, B. 1983. *América Latina en sus Lenguas Indígenas*. Caracas: Monte Avila Editores.
Silva, S A da. 1997. *Costurando Sonhos: Trajetória de um Grupo de Imigrantes Bolivianos que Trabalham no Ramo da Costura em São Paulo*. São Paulo: Paulinas.
Silva, S. A. da. 1999. Estigma e mobilidade: o imigrante boliviano nas confecções de São Paulo. *Revista Brasileira de Estudos de População (RBEP)*,16(1/2), Jan/Dec.
Silva, S. A. da. 2005. A praça é nossa. Faces do preconceito num bairro paulistano. *Travessia – Revista do Migrante*, São Paulo: Centro de Estudos Migratórios, No.51.
Silva, S. A. da. 2006. Bolivianos em São Paulo: entre o sonho e a realidade. *Revista de Estudos Avançados do Instituto de Estudos avançados da Universidade de São Paulo*. 20(57). São Paulo, May/Aug.
Vásquez, V H. 1991. *Historia de la Educación Boliviana, en Enciclopedia de la Educación Boliviana Franz*. La Paz: Tamayo, Teddy Libros Ediciones.

Chapter 5
Geography and Sustainable Development in Teaching and Education in Argentina*

Cláudia Marcela Polimeni

Introduction

The world is witnessing dramatic changes and undergoing processes of serious environmental deterioration and pollution. The irrational exploitation of natural resources has generated restlessness and concern for future generations and has forced scientists and politicians all over the world to include topics such as the environment, sustainable development and education in the agendas of international meetings and conferences held to protect the environment and address environment-related issues. At first, discussions were focused on uncontrolled population growth resulting in food scarcity, increasing urbanisation, shortage of drinking water supplies, indiscriminate deforestation, soil erosion, the advance of desertification, and the use of fossil fuel, that is, a great number of problems that needed to be studied and solved. Many of the resources we use are non-renewable and, therefore, a rational exploitation is required. The conflicting and changing world in which we live needs more than ever education and the contribution of different disciplines, including geography, in order to achieve sustainable development. Technology, as well as communications, is indispensable today to deal with this myriad of complex processes,

> The United Nations Decade of Education for Sustainable Development aims to promote education as the basis of a more varied society for humanity and to integrate sustainable development into the system of formal education at all levels. The Decade will also intensify international cooperation in favour of the common implementation of practices, policies and innovative programmes for sustainable development (UNESCO 2009).

---

*Translated by Prof. Mirta Rena de Kahn. School of Philosophy and Literature, University National of Cuyo.

C.M. Polimeni (✉)
University Nacional of Cuyo, San Rafael, Mendoza, Argentina
e-mail: marcela.polimeni@gmail.com

M.L. de Amorim Soares and L. Petarnella (eds.), *Schooling for Sustainable Development in South America*, Schooling for Sustainable Development 2,
DOI 10.1007/978-94-007-1754-1_5, © Springer Science+Business Media B.V. 2011

As has been clearly expressed in this UNESCO document, the Decade attempts to search for and to achieve the integration of a more supportive society through education. This can be accomplished by means of attitude and values changes and responsible decision making that contribute to improvements in the quality of life.

Background

A series of international meetings held to discuss the exploitation of natural resources and environmental protection provided concepts that have to be integrated into the curriculum design of school geography, not only in Argentina but also in the world. The environment has its own history and the outline below tries to reconstruct the facts that gave rise to the concept of sustainable development.

Over several decades, a series of historical events around the world resulted in the treatment of sustainable development in geographical education in Argentina:

1948 Universal Declaration of Human Rights.
1972 Declaration of the UN in Stockholm: "Man has the fundamental right to freedom, equality and adequate conditions of life, in an environment of a quality that permits a life of dignity and well-being, and he bears a solemn responsibility to protect and improve the environment for present and future generations" (Principle 1).
1975 Environmental Education (EE) is included in political agendas.
1977 Tbilisi Statement. Environmental education, its concept. Solving of environmental problems.
1980 The idea of sustainable development originating in Stockholm in 1972 is reinforced (UICIN/UNEP/WWF).
1983 World Commission for Environment and Development (WCED).
1987 Application of the principles stated in 1977.
1988 Commission on Environment and Development. Brundtland Report. Current concept of sustainable development defined.
1990 Jomtien Declaration on Education for All (UNESCO)
1991 Strategies for the future of life.
1992 Caring for the Earth: concrete actions and instruments for sustainable development.
2002 Johannesburg World Summit on Sustainable Development reinforcing commitment to support sustainable development.

The declaration of the Decade of Education for Sustainable Development had its background in these events that signalled the destiny of humanity. It is important to consider the concept expressed by Frers concerning what education for sustainable development means,

> ...more than just restricting it to a particular aspect of the educational process, it must become a privileged basis to develop a new life style. It has to be seen as an educational practice open to social life, involving members of the society according to their possibilities, in the supportive yet complex task of improving the relationship between humanity and their environment... (Frers 2008: 2)

The degradation of the environment continues endlessly and everyone is responsible for it. Industrial pollution is, indeed, a serious problem and when we observe our cities we realise that no concrete actions are being taken to alleviate environmental degradation. Despite the great amount of legislation concerning the limits to the degradation process, including environmental education, there is still a lack of legal application mechanisms. Great efforts have been made worldwide, yet awareness as to the need to educate for sustainable development is not significant. Argentina has serious problems related to environmental education and very few provincial governments carry out defined, concrete, comprehensive environmental policies, as is shown throughout this chapter.

Brief Review of the Existing Structure of Education in Argentina

Law No. 24.195, passed on April 29, 1993, has regulated the Argentine educational system. 2010 was a year of transition and changes in primary, middle and secondary schools are being planned. The present structure were based on:

1. Nine-year free and compulsory education, including kindergarten, primary and secondary schools (Kindergarten, EGB 1, EGB 2, EGB 3)
2. Three years of further education (Polimodal), for those over the age of 16. Three different kinds of orientations:
   a. Modalities (Natural Sciences, Economics and Organisation Management, Humanities and Social Sciences, Goods Production and Services, Communication, Arts and Design, among the most important). They ensure a general education focused on the core competencies of Geography I.
   b. Particular to each Modality (specific content for each orientation)
   c. Institutional Definition (EDI). Institutional context requirements and characteristics, demands, students' interests and needs are taken into account.

Teachers did not agree with the implementation of these educational structural changes. When forced to choose a Modality, confused parents and puzzled adolescents did not like Humanities and Social Sciences very much. On the other hand, Economics and Organisation Management was well considered since its actions were better defined.

The Teaching of Geography in Argentine Schools

The changes introduced by the Federal Law (1990s until now) not only transformed the structure of teaching in general, but also brought about substantive modifications for geography, some of them not particularly beneficial for the discipline, the teachers or the students. Professionals in the field were open to changes but high exposure to interdisciplinary work resulted in the weakening of specific content. While new content necessary to meet the dramatic changes happening in the world (economic, political and social) needed to be included, some others, just as important and basic, were neglected. They were no longer dealt with in geography but discussed from the points of view of other disciplines, for instance, biology. Also, the new epistemological approaches in geography which specifically aim at social issues put aside all the knowledge related to the environment and the natural context.

From the educational viewpoint, significant changes occurred: the number of hours devoted to the discipline resulted in a decrease in its importance in the general curriculum, and the subsequent loss of teachers' jobs had quite an important social impact on the fluctuating Argentine economy. Qualified teachers were assigned to other areas for which they had not been trained or to subjects that pretended to be geography. Concepts drawn from other areas such as sociology or economics were included, and the name geography was changed into Construction of Space, a concept borrowed from a different discipline. The subjects could be taught by professionals trained in other areas, some of them specialist teachers, others not. The situation led to discontent, confusion, anxiety, and vagueness in teaching, and the discipline was permanently confused with other areas of knowledge.

Impact on Content

The impact on content was considerable. While some topics were substituted, some new topics, also known as transdisciplinary or cross-curricular themes, emerged. In this respect, Benejam and Pages argues that,

> The absence of a unique disciplinary reference for the teaching of Social Sciences leads to an attempt to work out transdisciplinary concepts common to all the Social Sciences, which refer to the particular contribution of each area and that, as a whole, tell the students about the realities of the world they live in as well as its problems (Benejam and Pages 1998: 71).

Among those concepts, we can consider for environmental education the rational use of resources, sustainable development, globalisation, culture, poverty, marginality, natural risks and catastrophes, vulnerability. The incorporation of new technologies such as GIS and remote sensing are basic techniques for a renewed and more modern geographical education.

There was also a weakening in the geography content related to national identity and those topics referring to the physical-environmental area. Although new concepts were developed in the geographical subjects they were far from being considered part of geography. The compulsory application of content can be considered as a remarkable contribution: *conceptual* (region, geographical space, and landscape would be the traditional topics that cannot be omitted), *procedural* (cartographic elaboration and interpretation, reading of graphs), and *attitudinal* (where environmental education is included, sustainable development being one of its cross-curricular themes).

The Federal Law did not solve the problems observed in education.

> The NAPs (*Núcleos de Aprendizaje Prioritario* – Learning Priority Units) were designed by technicians and political representatives from Argentine provinces. The sense of its creation responded to the idea of unifying the national education system in order to bring about equality of knowledge and develop values conducive to the common good, social coexistence, shared work and respect for others. The NAPs constitute a set of core, relevant and meaningful knowledge that develops and expands students' cognitive abilities while catering for particular rhythms and learning styles through the creation of varied learning environments and conditions (Ministry of Education 2005: 68).

The NAPs should have unified content knowledge since the CBC (*Contenidos Básicos Comunes* – Common Basic Content) did not allow students to give evidence of their learning of the same quality. NAPs were not Learning Credits either. In spite of efforts to overcome the weaknesses of the Federal Law, the amendments were not sufficient to improve educational outcomes and a new Education Law foreseeing the combination of the traditional structure and Federal Law N° 25194 needed to be passed. The problem does not only refer to content but also to teaching practices: there are as many different teaching strategies as teachers who are also different as far as interests, values and experience are concerned.

National Strategy for Sustainable Development

Within the framework of cross-curricular or transdisciplinary content environmental education plays a crucial role, sustainable development being a key factor in the teaching-learning process. Its inclusion responds to the need to raise awareness of the use of natural resources among the present population and future generations. Such knowledge is highly important since the environment represents several problems, some caused by nature, many by humans. Some of the anthropomorphic causes include demographic pressure and industrial and uncontrolled urban growth. Other human activities resulting in natural resources degradation, such as mining and the constant deterioration of the environment, imposed the need to implement government policies to encourage sustainable development at both world and national levels.

What Strategies Did Argentina Implement?

The National Board for Sustainable Development was created in December 2010 and a basic document including its conceptual framework and the definition of three core principles was issued. Such principles are:

1. Healthy economy
2. Social Equity
3. Environmental Quality

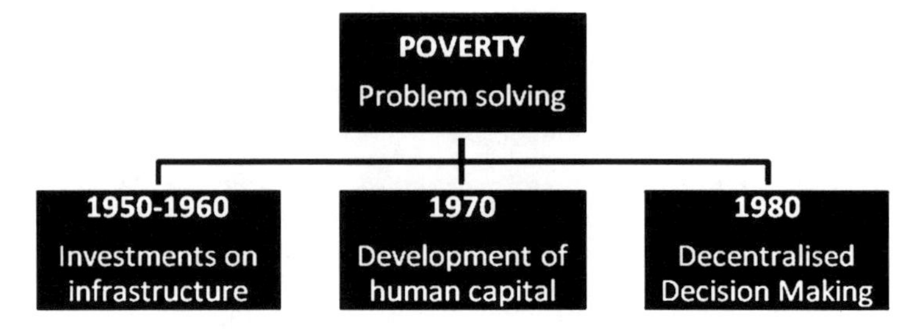

Fig. 5.1 Historical aspects of the attempt to resolve poverty-related problems. From 1950 to 1960 important investment in infrastructure was made. In 1970 the development of human capital was emphasised, and in 1980 decentralised decision-taking was addressed

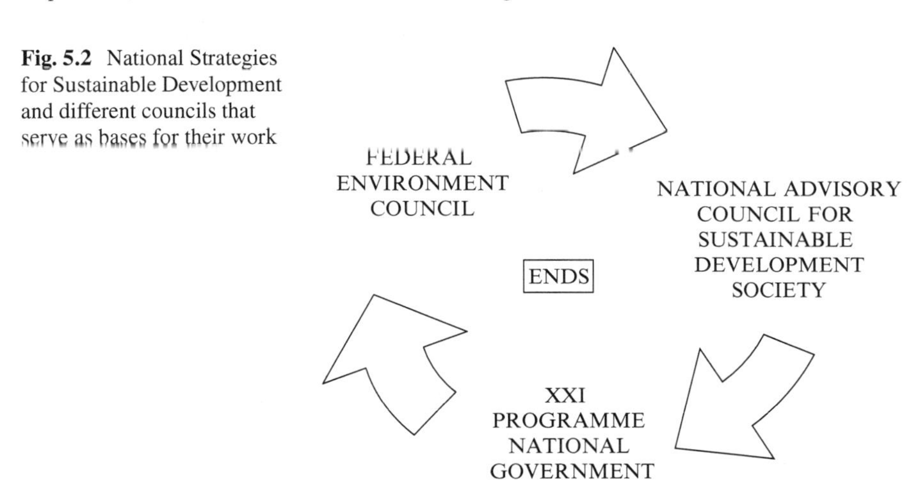

Fig. 5.2 National Strategies for Sustainable Development and different councils that serve as bases for their work

The goals emphasise social aspects and the economic growth which guarantees the quality, quantity and availability of natural resources for present and future development. The fight against poverty and the National Strategy for Sustainable Development (Estrategia Nacional para el Desarrollo Sustentable – ENDS) address sustainable development as a premise. A basic requirement in that fight entails education for sustainable development (Merenson 2001: 10). Social concerns shall be as integrated as economic aims. Economic growth must ensure the quality, quantity and availability of natural resources for present and future development.

In the evolution of the solving of poverty-related issues the 1970s stand out closely linking them to education. As can be seen in Fig. 5.1, Argentina implemented different strategies to solve such problems over the last 50 years.

At present, strategies are based on the concepts of 'transformation of society' and 'sustainable support', that is, an efficient approach to sustainable development through the fight against poverty. The environmental area is considered as a potential employer either in the water market, the waste market and/or clean development mechanisms. Among the goals of ENDS, planning, participation and information, and implementation of national indicators for sustainable development stand out. In order to achieve these goals, it was necessary to create the Federal Environment Council, the National Advisory Council for Sustainable Development, to identify and implement local and regional points in the XXI Programme. It is observed that environmental education (EE) is not considered within the new strategies (See Fig. 5.2).

Education-Related Policies and National Legislation to Achieve Sustainable Development

Within the framework of education for sustainable development, national laws and policies are fundamental to realise and implement 'sustainable development'.[1] Some instruments are not entirely appropriate for this purpose, though the General Environment Law N° 25675 (passed by the Argentine Senate and the National Congress on November 6, 2002) is considered as a starting point.

Article 1 establishes a series of bare minimum budgets for the achievement of sustainable management which should be appropriate to the environment, the preservation and protection of biological diversity, and the implementation of sustainable development.

Article 2 states that the Environmental Policy shall meet 11 goals, one related to environmental education (EE), which should include education for sustainable development. Regarding the ten principles set out by the Environmental Policy, only one considers and maintains sustainability as a basis for economic and social development and exploitation of natural resources.

Among the "policies and environmental management instruments", six principles are established in Article 8, but only one of them is aimed at environmental education.

Article 14 determines the basic principles for generating citizen values, behaviours and attitudes consistent with a balanced environment, aiming at the preservation of natural resources and their sustainable use to improve the quality of the life of the people.

Article 15 considers that "...environmental education will be a continuous and permanent process resulting from the articulation of the various disciplines and educational experiences, and which shall facilitate the comprehensive perception of the environment and the development of environmental awareness". Formal education is certainly important for the consolidation of sustainable development since it emphasises the cross-curricular characteristics of the theme with an interdisciplinary, multidisciplinary, and disciplinary approach. It is not only immersed in geographical knowledge but goes beyond the discipline as well (General Environment Law N° 25675).

[1] The concept of sustainable development is taken from the Brundtland Commission: "sustainable development is the development that meets the needs of the present without compromising the ability of the future generations to meet their own needs" (CMMAD 1987: 42–187).

It is also demonstrated that the competent authorities have to coordinate with the Federal Councils for the Environment (Consejos Federales del Medio Ambiente – COFEMA – created in 1990) and with the Council of Culture and Education to carry out plans and programmes in formal and non-formal education systems. It is the responsibility of the different jurisdictions to implement their own programmes and curricula according to the relevant standards.

A joint resolution, No. 178/2007, was issued and through it an Advisory Committee for Financing Environmental Management was created, its main objective being to advise the authorities on the implementation of General Environment Law N° 25675, approved on February 19, 2007.

The 'constitutionalisation' of environmental education (Law 24.309) included an environmental clause in the new text (Article 41) stating that "...the duty of the authorities is to provide environmental education" (for the contributions of environmental sustainability and sustainable development to the debate on the New Law of Education). It makes reference to the General Environment Law and states that environmental education is part of the public order, both as an objective and as an instrument. By General Environment Law N° 25675, sustainable development is identified as a state policy in Argentina and it is to be achieved through environmental education (EE).

Sustainable development includes three components: environment, society, and economy. Considering the New Law of Education that will begin to be implemented in the coming years, environmental education is, "...the basis for achieving a fairer and more sustainable society...it is the key to sustainable development" (Secretary of Environment and Sustainable Development 2008a: 25–42).

The document ends with a set of recommendations, among them, the incorporation of environmental education (conceptually, procedurally, and attitudinally) within the Common Curricular Content for all the jurisdictions and the priority knowledge units. This content has to be learned by all the students in the country, at all levels within the national educational system. The document also lists all the content needed to achieve environmental education and emphasises sustainable development. Indeed, there is a legal framework for environmental education and the beginning of the implementation of sustainable development.

Education for Sustainable Development in Argentina

In the course of the country's institutional life, different Secretariats and/or Ministries that were part of the national government have been created and dissolved. The political organisation of the current government includes the National Secretariat for the Environment and Sustainable Development which coordinates three under-secretariats, each one responsible for environmental control and pollution prevention.

- Under-Secretariat for Planning and Environmental Policy.
- Under-Secretariat for Promotion of Sustainable Development: National Board for Sustainable Development Management.
- Under-Secretariat for the Coordination of Environmental Policies.

The Under-Secretariat for Promotion of Sustainable Development is responsible for assisting the Secretary in the formulation and implementation of a national sustainable development policy. Second in importance, according to the objectives of this chapter, is the Under-Secretariat for Planning and Environmental Policy, which assists the Secretary in the design and implementation of national policies related to the rational use of natural resources, conservation of biodiversity, development of instruments and implementation of policies related to social, economic, and environmental sustainability. This Under-Secretariat is also responsible for designing strategies at regional level and for other areas of the environment and sustainable development.

Undoubtedly, the intention of recent governments has been based on the activation of sustainable development (as we have seen, there is a structure for success) but there are as yet no concrete actions, especially in the definition of an education policy based on sustainable development. An environmental education (EE) area was developed, established as a fundamental component aimed at finding alternative paths that enable the construction of a different society that is fair, participative and diverse. Through the Environmental Programme Coordination Unit,

> ...it is sought to discuss the meaning of educational and environmental processes undertaken from several municipal and provincial areas, as well as to promote educational and environmental practices as critical discussion spaces to encourage and support the process of changes aiming at development (Secretary of Environment and Sustainable Development 2008b: 7)

Within the National Strategy for Environmental Education another area was created, "...with the purpose of defining the National Policies for Environmental Education guidelines, and setting and achieving actions based on a deep knowledge of the reality of EE in Argentina" (p.4).

One of the formulated goals, which is relevant to this chapter, is to "...create and maintain spaces of coordination and articulation with other sectors of public administration, especially national and provincial ministries and secretariats of education" (*idem.*). It also created another dependent entity in this area, "National Strategy for Environmental Education to set guidelines of the National Environmental Education Policy to set and achieve their actions on the basis of sufficient knowledge of the reality of the EE in Argentina" (p.7). Among the objectives raised, and what is relevant for this chapter, is the statement to, "...create and sustain spaces of articulation and coordination with other sectors of public administration, education ministries and secretariats both national and provincial" (Secretary of Environment and Sustainable Development 2008b: 2).

In this sense, a thorough analysis of the CBC (Common Basic Contents) in the EGB and Polimodal levels of formal education in Argentina has to be in order to identify the articulation needed to develop EE. In the Secretariat website (http: //www.ambiente.gov.ar) there are a number of documents related to the environment, an environmental glossary, statistical indicators, etc., which are particularly interesting and very important when teaching EE at all levels of formal education in Argentina. It also includes more than 20 projects and programmes on the environment and sustainable development.

Sustainable Development in the Common Basic Content and in Argentine Education in General

The General Law for the Environment N° 25675 states that both environmental education and protection of the environment are to be considered state policies. In 2007 the national government created the National Research Institute for Sustainable Development. Daniel Filmus, then Minister of Education, emphasised the leading role of the state regarding sustainable development. He underlined also the necessity to create synergy between all projects and programmes in progress under state supervision. First, sustainability was declared to be a state policy for the period 2010–2016. Secondly, the Institute was appointed as the regulating agency responsible for determining guidelines and, thirdly, the Institute was given responsibility for promoting education for sustainable development.

The IADS (Instituto Argentino para el Desarrollo Sostenible – National Research Institute for Sustainable Development) aims to:

(a) Generate innovative proposals to achieve sustainable development through the promotion, study and implementation of practical and research projects and programmes.
(b) Contribute to the design, development and implementation of public policies for sustainable development.
(c) Establish cultural and scientific exchange programmes with similar national and foreign agencies.
(d) Encourage sustainable production and consumption patterns in society, and in production and government sectors.

Its main objectives are to:

(a) Encourage projects for diagnosis, implementation, monitoring and assessment of sustainable development related works.
(b) Design a Discussion Board to disseminate and promote the themes developed.
(c) Organise, present and participate in courses, workshops, seminars and any other training meetings related to the objectives.
(d) Participate and/or take part in the development, implementation and assessment of projects developed by other organisations, associations, public or private agencies with similar aims (IADS 2008: 4)

On the one hand, the scope of action highlights education and training as initiatives to help consumers and producers to adopt sustainable consumption and production practices through knowledge and capacity-building. On the other hand, and within the framework of non-formal education, the OSC's (Organizaciones de la Sociedad Civil – Civil Society Agencies) should be mentioned. They are numerous and play a very important role in the education of the general public. One of them, perhaps the most well-known for its work and background, is ECOPIBES, (http://www.ecopibes.com.ar). ECOPIBES is an educational project created to encourage sustainable development among children and youngsters. This public agency, founded

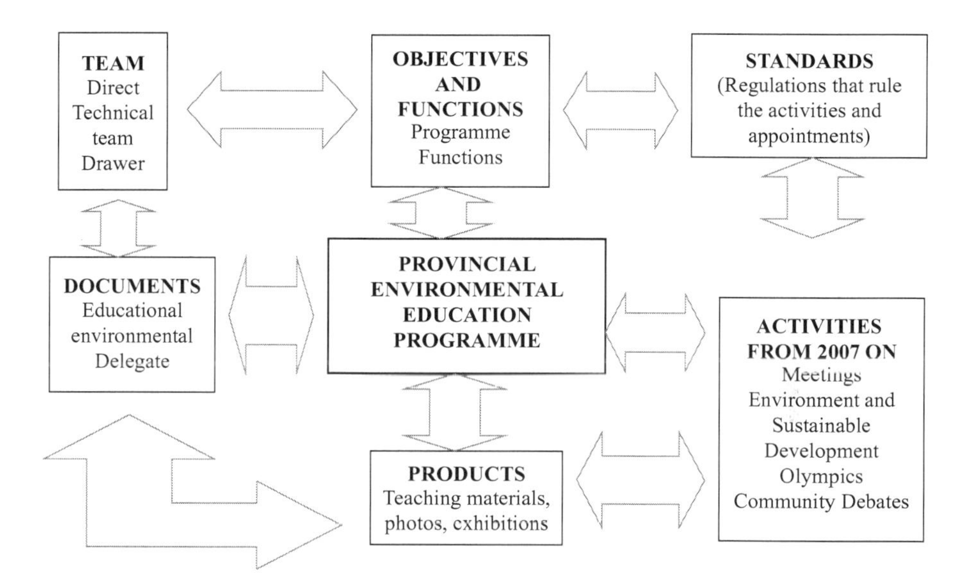

Fig. 5.3 Programme of the Province of Entre Ríos environmental education outline, tailor-made to cater for local needs. It was implemented by the Provincial Secretariat of Environment and Sustainable Development. Since 2007, its activities include the Environment and Sustainable Development Olympics. It is an example of non-formal education

ten years ago, is an environmental network whose goal is, "to offer an effective means of dissemination, education and networking for people committed to sustainable development, regardless of their group affiliation or origin" (ECOPIBES 2010: 9). It is supported by the UNEP and the National Secretariat for Environment and Sustainable Development. On its website (www.ecopibes.com) there are all kinds of icons to learn about the environment. Sustainable development is one of the fields covered and it includes an explanation of its origin and components and the bases for what a sustainable society should be like, and the roles children should play in it are discussed. It also includes a space for the views and comments of young people. Different game-like strategies for teaching environmental protection are used, including a series of highly pedagogical e-books from the UNEP (*GEO-Youth for Latin America and the Caribbean*) that deal with many interesting themes.

In the field of *non-formal education* the province of Entre Ríos should be noted as an example. Its government implemented a policy for environmental education (http://www.entrerios.gov.ar/CGE/index) with a rational and unique programme that is synthesised in Fig. 5.3 and is called "Environmental Education Programme of the General Council of Education" (http://www.entrerios.gov.ar/CGE/index, Accessed: 26 July 2009).

The implementation of the Plan and its continuity is as important as the intended purposes. Among its activities, the Environment and Sustainable Development Olympics, held for 3 years in a row, are worth mentioning. They have made significant progress for the province concerning environmental education and they have also taken important steps to raise awareness about sustainable development.

What Importance Is Given to Sustainable Development in National and Provincial Curricula?

At the end of the 1990s, thorough changes were introduced in formal education in Argentina and geography was deeply affected by them. Unlike other disciplines, geography was deprived of much of its significant formal content. Its name was changed, its identity lost, and its content blurred and impoverished. Cross-curricular themes became the new and attractive centre of attention and their emergence was accompanied by different historical, political and social processes among which globalisation, sustainable development, environment, violence, health, and race, can be mentioned. They turned out to be a novelty in the process of modernisation of the educational system. Although the concept of sustainable development gained importance in geography classes, it was absent in most of the Argentine provincial curricular designs. Only some jurisdictions made it explicit, e.g. La Pampa, Santa Fe, Buenos Aires, Neuquén and Santa Cruz, which included it in different curriculum areas, axes, levels and years, as shown in Fig. 5.4.

PROVINCE	CURRICULAR AREA	AXES	LEVEL	YEAR
La Pampa	Argentine and Mercosur Geography	Man and environment	Polimodal	2°
Neuquén	Argentine and Latin American Geography	Societies and geographic space	EGB3	2°
Santa Cruz	Geography II	Environment and Sustainable Development	Polimodal	2°
Buenos Aires	Argentine Geography	Environmental Issues	Polimodal	2°
Santa Fe	Geography	Unit 1	EGB3	1°-2°
Mendoza	World Geography	Environmental Issues	Polimodal	1°
	Regional Geography	Units I and II	Polimodal	1°
Misiones	General Geography	Environmental Issues	Polimodal	1°

Fig. 5.4 Sustainable development content in the curricula of Argentine schools. Tailor-made on the basis of jurisdictional curriculum designs. There is an obvious lack of uniformity with the same content not always in the same level, year and/or curricular area

The lack of information and non-appearance of the concept in curricular design have not been due to the lack of governmental policies. Environmental policies are indeed public concerns. There is an administrative and legal framework. However, there are no mechanisms that coordinate national government plans with educational decisions. The creation of NAPs, mentioned above, has been an attempt to solve the problems arising from the implementation of the Federal Law and to unify Argentine education. But NAPs also failed and teaching sustainable development was not considered as a goal. A new Law of Education is about to be implemented, but then again geography is not a priority; further weakening of the discipline is anticipated.

Conclusion

There are several important difficulties confronting the implementation of policies related to environmental education and sustainable development in Argentine formal education:

1. There is no coordination between the government agencies and the national and provincial ministries of education to implement policies for environmental education. A common strategy developed jointly by the Federal Government, the Secretariat of Environment and Sustainable Development and the Ministry of Education has not been planned, which would explain the absence of environmental education in curriculum design.
2. Sustainable development-related content is not present in all the provincial curricular designs. In fact, there are very few provinces whose ministries are committed to and recognise the importance of its teaching in all levels of the educational system. Therefore, the topic has been left to the geography teachers' goodwill. The failure of the NAPs needs to be mentioned again since they did not achieve the expected results, as well as the weakening effect that the implementation of the Federal Law had on Argentine education in general and on some geography content such as sustainable development.
3. There are no plans that take into account teachers' training in sustainable development, thus its implementation in classrooms is now just an ideal.
4. There is a lack of centralised programmes planned by the National Government to be implemented by the provincial governments.
5. Teachers' training and updating. It is necessary to design and use more modern and dynamic teaching-learning strategies, appropriate to sustainable development to help students acquire the discussed topics so that they can be translated into practice.
6. There is insufficient technical preparation and a lack of technological support to implement a more modern education suitable to current times.
7. General Environment Law N° 25675 established a minimum budget for the implementation of sustainable development. Article 8, Paragraph 4, defines environmental

education as one of the instruments of environmental policy and management. In the same article, Paragraph 6 determines the economic arrangements for the promotion of sustainable development and environmental management. The latter emphasises that national government and the territories will encourage inter-jurisdictional coordination between the provinces and municipalities through the Federal Environment Council (COFEMA – Consejo Federal de Medio Ambiente). Research results indicate that the Council does not work properly and that there are not enough mechanisms to implement that coordination.
8. Sustainable development content is absent in most provincial curricular proposals and, when included, it is not dealt with as it should be according to what is expressed in Law No. 25675 (Articles 9, 14 and 15) concerning environmental education, the central axis where sustainable development is anchored. The Law needs to be fully implemented. The coordination between the national and provincial governments has to be started in the short term so as to include sustainable development as a compulsory content at all levels. It is also necessary to encourage the formation of spaces for the participation of the community to discuss and stimulate education for sustainable development.
9. There is a need to generate activities to promote education for sustainable development in non-formal education to raise awareness and build more solid foundations for sustainable development. Some efforts made from non-formal and formal education are worth mentioning. The Government of Entre Ríos has a Comprehensive Environmental Education Programme that every year carries out several activities such as the already mentioned Olympics.

In conclusion, new government coordination strategies are required; both the environment and environmental education have to be considered in state policy as does sustainable development. Resources are non-renewable and the lack of awareness and environmental values put all the plans for the building up of sustainable development in danger.

References

Benejam, P, and J Pages (eds.). 1998. *Enseñar y Aprender Ciencias Sociales, Geografía e Historia en la Escuela Secundaria (Teaching and learning Social Sciences, Geography and History in secondary schools)*, 2nd ed, 71–94. Barcelona: ICE, Barcelona University.

Comisión Mundial del Ambiente y Desarrollo. 1987. *Nuestro Futuro Común*. Madrid: Alianza Ed.

ECOPIBES. 2010. Desarrollo Sustentable Ambiental, Social y Economico Para Todos. http://www.ecopibes.com/mas/ desarrollo. Accessed 12 Feb 2010.

Frers, C. 2008. Promoviendo una Educación Hacia el Desarrollo Sostenible. http://www.ecoportal.net. Accessed 31 May 2008.

IADS–Instituto Argentino para el Desarrollo Sustentable. 2008. http: //www.iadsargentina.org. Accessed 25 Jan 2009.

Merenson, C. 2001. *Estrategia Nacional de Desarrollo Sustentable* (National Strategy for Sustainable Development). Basis Document, National Board for sustainable Development: 10.

Ministry of Education. 2005. Núcleos de Aprendizajes Prioritarios. www.me.gov.ar/curriform/nap.html. Acessed 12 July 2009.

Secretary of Environment and Sustainable Development. 2008a. Ley de Educación Nacional. http://www.Debate-Education/ debate-educacion.educ.ar/ley/. Accessed 10 Feb 2010.
Secretary of Environment and Sustainable Development. 2008b. Estrategia Nacional de Educación Ambiental. http://www.ambiente.gov.ar/default.asp?idseccion=231. Accessed 07 Feb. 2010.
UNESCO. 2009. Tendencias de la Educación Ambiental a partir de la Conferencia de Tbilisi. http: //www.unesco.org/es. Accessed 12 Aug 2009.

Chapter 6
Venezuela and Education Transformation for the Development of the People

Rosa López de D'Amico, Maritza Loreto, and Orlando Mendoza

The first duty of a government is to give education to the people

Simon Bolivar (1992)[1]

Introduction

To write about education in Venezuela in only one chapter is indeed challenging, maybe even more challenging than the huge changes that have occurred in the education system especially in the last ten years. Before 1999, there were changes marked by the country's historical development (López de D'Amico and González 2006), nevertheless the changes that have taken place in the last ten years start with the constitution and national education legislation. Fortunately we can say that educational changes have been made for the benefit of Venezuelan society. Until recently, when people referred to Venezuela they tended to associate it with oil and beauty contests. However, due to the mass media there are now two diverse views, either it is seen as a revolutionary vibrant country or as a place where there is social tension, it all depends on the view of those who control the

[1] Simón Bolívar (1783–1830) is the most important Venezuelan national hero and the precursor of independence in the Latin American continent. He is regarded in Latin America as a hero, visionary, revolutionary and liberator. During his short life he led Bolivia, Colombia, Ecuador, Panama, Peru and Venezuela to independence and laid the foundations of Latin American ideology on democracy.

R.L. de D'Amico (✉)
Research Centre for Studies in Physical Education, Health, Sports, Recreation and Dance, Pedagogical University Experimental Libertador (UPEL), Maracay, Venezuela
e-mail: damicolopez@cantv.net

M. Loreto • O. Mendoza
Pedagogical University Experimental Libertador (UPEL), Maracay, Venezuela

M.L. de Amorim Soares and L. Petarnella (eds.), *Schooling for Sustainable Development in South America*, Schooling for Sustainable Development 2,
DOI 10.1007/978-94-007-1754-1_6,

information. But what is clear is that we have the oldest democracy in Latin America and currently it is the place where people can express their ideas openly and where changes are the products of democratic elections. These can be sustained by those who live there and have experienced the social changes in the last fifty years or more.

The Constitution of the Bolivarian Republic of Venezuela, adopted in 1999 after consulting the citizens of this country, comprehensively addresses the type of education the population should receive. Its conceptual structure is based on: access to knowledge and information, cooperation as a means of participation and involvement of citizens, social economy, public security, science and technology, responsibility of all citizens in public affairs, endogenous and sustainable development, employment, inclusion and social justice, social security, social control, participative democracy, responsibility, communication, popular sovereignty and missions. It also includes international aspects related to the multi-polarity of international society, Latin American and Caribbean integration, security and national defence.

In the particular case of the social missions, relating to the area of education, the Venezuelan state has implemented the Mission Robinson I, Mission Robinson II, Rivas Mission, the Sucre Mission, that together with the Science and Culture Missions complement what constitutes the entire educational system, designed to combat exclusion and promote participation, inclusion and reinsertion of Venezuelans who at some point have dropped out of the education system. Mission Robinson I is aimed at combatting illiteracy. Mission Robinson II is aimed at people who have learned to read and write and want to continue to study and achieve the sixth grade (year 6). Rivas Mission serves peoples who were excluded from school after they have achieved sixth grade, so it is desirable that they complete high school. Finally, the Sucre Mission seeks to provide an opportunity for students who have finished high school, and could not enter universities in the past, to continue their studies so that they could attain a college degree. These educational missions constitute a complementary system of education that the state provides to include those people who for various reasons have not achieved a formal education, particularly in the past. In recent years, these programmes have given opportunities for many to have access to education.

The state also has its traditional education system or mainstream education system, consisting of pre-schools, elementary (primary) schools, high schools, technical colleges and universities. They develop the programmes for the Bolivarian Education: Simoncito (Little Simon) for children younger than 6 years; Bolivarian Schools (grades 1–6), High Schools: (years 1–5 or 6) or Robinsonian Technical High Schools (years 1–6) and University. This structure of the Venezuelan educational system has a philosophical underpinning in the Constitution of the Bolivarian Republic of Venezuela under the Latin American ideal and endogenous and sustainable development.

Taking into account previous considerations, we discuss whether it is possible to evaluate an educational system with these characteristics using quantitative evaluation reference schemes that do not consider within their own parameters of measurement, integration, participation or values and focus attention on performance and the structural conditions of the system. The products of the education system cannot be assessed only from a quantitative point of view, in terms of mastery of learning or achievement and accomplishment, or material conditions of the institutions that provide education for the population. Particularly because Venezuelan education is based on humanism and solidarity, priority over material conditions, citizenship, responsibility and participation are key elements of the constitutional principles.

The present chapter is divided into four sections; it starts with the introduction, then continues with a short account of the experience in Aragua state, then continues with a review of the constitution, later the focus is on the education missions in place and finally the chapter ends up with discussion and conclusions.

A Glance at Aragua State Experience

... either we invent or we err

Simon Rodriguez (Rumazo 2008)[2]

Loreto (2004) points out that it was from 1991 that discussions began in Aragua state about the possibility of achieving the beginning of a transformative process through the state apparatus. It started with proposals from the Ministry of Education. In 'Barbacoa', teachers, parents and representatives discussed many things together, including the exclusive nature of education particularly in the southern towns of Aragua state, where poverty and inequality were common elements. They participated in the construction of the Aragua Educational Project in which the basic aim was to achieve the democratic participation of the people (teachers, students, parents, and communities). At the same time, important political changes were occurring in the country, such as the elections of state governors and the creation of new structures for state institutions, and new possibilities for reaching the basic structure of the school system. It was understood that there were not many opportunities for changes due to the policies imposed on Latin America for decentralisation of educational programmes financed through the World Bank (WB) – International Monetary Fund (IMF) – including teacher training, school physical infrastructure, teachers' support centres, and libraries. There was no reference to the curriculum which, basically, had to be evaluated and transformed. Nevertheless, the situation was used to set up working groups using action research to analyse aspects of regional education. Significant progress was made primarily on a new vision of school management and teacher training. The IMF's prescription for Latin

[2]Simón Rodríguez (Caracas, Venezuela, 1769 – Perú, 1854), known during his exile from Spanish America as Samuel Robinson, was a South American philosopher and educator, notably Simón Bolívar's tutor and mentor.

American education focused on strengthening the Basic Education programme which began in the 1980s and was reviewed in 1998. It included a curriculum design for the first and second stages of basic education, with the implementation of transversal axes, planning classroom educational projects and pedagogical projects. These were introduced despite the fact that since 1992 the State Education Office (SEO) had implemented other projects and innovations which included a total 25 pilot schools. In other words, the IMF recommendations were not targeting the real problems existing in a type of education that could help to look at endogenous development and to target global problems such as environmental education, to mention but a few issues.

In 1997 the voices became stronger and people were speaking about the most important historical process in our country, because for the first time ever Venezuelans were experiencing participation in a constituent assembly to draft a new constitution and with it a new national project. A group of educators from around the country formed a National Front for Educational Transformation (Constituent). In 1998 ideas and proposals for educational change were written into a text titled *School for Life and Freedom*. This proposal was seen by the newly appointed Minister of Education (1999) who invited us to join his team (Loreto 2004).

The Ministry of Education started working with the SEO on the specific proposal and then the constitutional process of the National Constituent Education was formally initiated in the country. The debate took place in most schools of Aragua state, each school submitted its proposal about 'the country we have and the one we want', 'the teachers we have and what we want', teacher education, school infrastructure, educational partner programmes, socio-economic, medical and recreational services, union status, and other topics. There were intensive and extensive working days conducted at different levels: parish, municipality and regional level. These produced a first document entitled *Aragua Educational Proposal for the Construction and Systematisation of the National Education Project* which expressed for the first time the ideas, feelings and words of educators from the grassroots. In the past the education system had been designed by experts, in addition to being an almost exact copy of educational models from other countries. It is important to acknowledge the integration of SEO and the Ministry of Education sector in all conferences as well as the efforts made to coordinate education policies for the state. This document was delivered to the Ministry of Education and the representatives of Aragua state in the National Assembly. The National Education Project (PEN – 1st version) emerged in the systematisation of the proposals from all states. One of the central ideas of this methodology and the project itself is that social change is not imposed but is constructed collectively and permanently. At present, it is advancing with different degrees of development in various areas of the country on matters relating to curriculum, school and community, teacher training and teachers' working conditions, among others, encompassing all that constitutes an educational project.

The Ministry of Education was changed in 2002 to the Ministry of Education, Culture and Sports (MECD), having a Vice-Ministry of Educational Affairs, Sports,

Culture and Higher Education, the latter then became the Higher Education Ministry (this restructuring at first did not help as there remained many vices inherited from the old structure: disruption, patronage, disorganisation, poor communication, etc.). Officially, the MECD and the Educational Office in each state (SEO) were required to create a new institutional framework that brought power to the people, on the principles of shared responsibility and cooperation.

Education and the Constitution of the Bolivarian Republic of Venezuela

Without popular education there will be no true society

Simón Rodríguez (Rumazo 2008)

December 19th, 1999, marked the beginning of the new constitutional document that governs the fate of the Bolivarian Republic of Venezuela (Constitution of the Bolivarian Republic of Venezuela (CRBV)). From that moment a process in this country began aimed at reshaping the Republic through collective construction. The preamble states:

> ...a democratic, participatory and self-reliant, multiethnic and multicultural society in a just, federal and decentralised State that embodies the values of freedom, independence, peace, solidarity, the common good, the nation's territorial integrity, comity and the rule of law for this and future generations; guarantees the right to life, work, learning, education, social justice and equality, without discrimination or subordination of any kind... (CRBV 2006: 5).

The preamble to the Constitution of the Bolivarian Republic of Venezuela outlined the sort of citizen required to build a society in which social injustice is removed and equity is achieved (MECD 2002). However, even though these are the particular principles of the Venezuelan state, it is necessary in order to achieve the required transformation to review theoretical and philosophical concepts that underlie a model still under construction and may be contradictory to it because they belonged to the previous scheme.

Article 1 of the Constitution, states that,

> The Bolivarian Republic of Venezuela is irrevocably free and independent, basing its moral property and values of freedom equality, justice and international peace on the doctrine of Simón Bolívar, the Liberator (*ibid.*: 7).

Freedom must be understood beyond the concept of membership of a geographical space. It must be viewed from the standpoint of people's independence to act and make their own decisions without blackmail or business ties, political, cultural, ideological and military, to enable the achievement of sovereignty, immunity, territorial integrity and national self-determination.

Article 2 of the Constitution states that,

> Venezuela constitutes itself as a Democratic and Social State of Law and Justice, which holds as superior values of its legal order and actions those of life, liberty, justice, equality, solidarity, democracy, social responsibility and, in general, the preeminence of human rights, ethics and political pluralism (*ibid.*).

Article 3 is expressed as,

> …essential purposes of the State are the protection and development of the individual and respect for the dignity of the individual, the democratic exercise of the will of the people, the building of a just and peace-loving society, the furtherance of the prosperity and welfare of the people and the guaranteeing of the fulfillment of the principles, rights and duties established in this Constitution. Education and work are the fundamental processes for guaranteeing these purposes" (*ibid.*).

These first two articles of the CRBV initially point out the direction to follow to fulfill the purposes of the Venezuelan state. Knowledge of human rights and their defence, as well as personal development, peace and welfare of the people, can only be achieved through the sum of the consciousness of individuals belonging to the same community, and this is only possible through education and work.

Later in the CRBV, Article 102 establishes the compulsory and free characteristics of education; corresponding to Articles 2 and 3, because historically there were Venezuelans who did not have access to it. In the period between 1958 and 1999 education had been changing to become privatised, contributing to the exclusion of the disadvantaged. In Article 102 it is emphasised that, "The State assumes responsibility for it as an irrevocable function of the greatest interest, at all levels and in all modes, as an instrument of scientific, humanistic and technical knowledge at the service of society" (p.38). This ensures state involvement in the shaping of the society and, moreover, it gives education its rightful role in shaping the community. Likewise, the state assumes a leading role in education and confronts the neo-liberal doctrine of non-participation and exclusion in fundamental aspects such as, health, education, economics, national defence, sovereignty, self-determination of peoples and building a pluralistic and egalitarian society by citizens. Education is understood to be a 'public service', which gives it a character of special interest to every citizen. It becomes an object of participation "and is based on all currents of thought", thus giving it a character which embraces diversity and universality of all members of this multicultural society. This is complemented by the statement that declares that education has, "…the purpose of developing the creative potential of every human being and the full exercise of his/her personality in a democratic society based on valuing the work ethic and the active, conscious and joint participation in the processes of social change" (*ibid.*). This was visible and necessary from the time of the adoption of the Constitution and the project that was planted with it was imbued with the values of identity and a Latin American and universal vision. The participation of society is in the last lines of the Article where it is stated that the State with the participation of families and society promotes the process of civic education in accordance with the principles contained in the CRBV and the law.

Article 103 states that everyone has the right to a full, high-quality, ongoing education under conditions and circumstances of equality, subject only to such limitations as derive from the person's own aptitudes, vocation and aspirations. Education is compulsory from the maternal to the diversified secondary level. The State is obliged to serve the population in relation to education from conception and therefore to take care of the mother and family. Education is also free up to university undergraduate level and the state provides the level of investment as recommended by the United Nations Organisation. The state will create and maintain institutions and services sufficiently equipped to ensure access, permanence and completion in the education system. This constitutes a key element against social exclusion and school dropout. In this Article, to ensure inclusion of all, it indicates that the law shall guarantee equal attention to people with special needs or disabilities and those who are deprived of their freedom or lack of basic conditions for their incorporation and permanence in the educational system. Finally, in support of investment it establishes that the contribution from the private sectors to projects and programmes of public education will be recognised as tax deductible on income according to the pertinent law.

Education is conceived as a whole beyond the school environment and this is part of the curriculum, to demystify the idea that school is an isolated case of reality and activities outside the classroom or outside of schools are extracurricular. This is why it is conceived that culture and sport should be an integral part of any curriculum proposal. This includes all cultural, idiosyncratic and traditional events from all regions, as well as sport as an integral element of personal training. This is based on Articles 100, 111 and 107. Article 100 states:

> The folk cultures comprising the national identity of Venezuela enjoy special attention, with recognition of and respect for intercultural relations under the principle of equality of cultures. Incentives and inducements shall be provided for by law for persons, institutions and communities which promote, support, develop or finance cultural plans, programmes and activities within the country and Venezuelan culture abroad. The State guarantees cultural workers inclusion in the social security system to provide them with a dignified life, recognising the idiosyncrasies of cultural work, in accordance with law.

In terms of environmental rights there is a whole chapter in the CRBV dedicated to them but besides that the environmental issue is indicated in various Articles of the Constitution. Moreover, in terms of environmental education, Article 107, in particular, emphasises that environmental education is obligatory in the various levels and modes of the education system, as well as in informal civic education. The teaching of the Spanish (Castilian) language, history and Venezuelan geography and the principles of Bolivarian thought are mandatory in public and private institutions, up to the diversified cycle level. This is followed up, after much controversy, in the recently approved Organic Law of Education (LOE 2009) which clearly states the relevance of environmental education and includes it in the content of the curriculum.

Environmental education has been defined since 1979 by the International Union for Nature Conservation as a process, so in that sense it looks at promoting in individuals an adequate relationship with the environment. It is proposed that this action could be achieved acknowledging the values of the environment as well as building a whole body of knowledge that promotes environmental education and a literacy programme on environmental education. In that way the connection between knowledge and the values of education could support environmental education. This could not be achieved in Venezuela because environmental education was not even included as a strong component in the education system. The CRBV provided a solid base for including environmental education in the education system; it does indeed give a big push towards environmental education. Nevertheless, the changes at school did not occur because of the big debates that started against the constitution and the government itself. So the constitution was in place but the curriculum was the same as before 1999, until it started to be changed and content such as environmental education started to be included in the parallel projects that the government was developing (missions). Currently, it is clearly established in the LOE that environmental education is compulsory at all levels of the education system. Nevertheless, there is still a gap between what is established in the CRBV and LOE because contents are still developed in isolation and not integrated with other areas of the curriculum. More time is needed to implement and internalise all that is desired in the LOE.

Environmental education is directly related to sustainable development because the purpose of the latter is to propose a new way, through reflection and changing global actions, towards a new sort of development and to start thinking about what for, who for and how this development should be undertaken. It has to be acknowledged that there has been an effort to build a body of theory of environmental education and sustainable development as both look at an education directed towards a long-lasting and fair world for all kinds of living things.

In Venezuela there are projects being developed at universities, government institutions (in combination with the agendas of several ministries) and NGOs that promote and develop environmental themes according to the characteristics of the locality, involving schools and communities. An interesting project has been conducted by a research cluster in environmental education (NIAFE: Research Centre for the Environment and Specific Purpose) located at the UPEL (Universidad Pedagógica Experimental Libertador – Teachers Training University), which year after year approaches the same community and its students and works with the school children using various strategies to help them become aware of the environment in which they live and what they can do in order to have a better conscience about it. The concept of 'ecological schools' or 'green schools' also exists but so far these schools have been isolated experiences. Stronger emphasis has been given towards the denouncement of ecological damage that has been caused to the environment. At this stage there has been an awakening of interest in environmental education, but there is still a long way to go.

In terms of teachers, and teachers training, we have made a step forward in the sense that there have been graduate programmes created to train professionals in environmental education. In the past it could be the biology and/or geography teacher who would refer to environmental education in a very peripheral way. But this is starting to change, and in undergraduate programmes there are courses in environmental education. Environmental education as presented in all the programmes is closely associated with sustainable development as established in the CRBV.

Education and the National Economic and Social Welfare Development Plan (2001–2007)

A responsible State and with real authority assumes as its role the general orientation of education. This guidance expresses its political doctrine and therefore forms the conscience of citizens

Luis Beltrán Prieto Figueroa (2002)[3]

The constitutional framework of Venezuelan education is implemented and made effective through the Plan for Economic Development and Social Welfare (PDESN 2001). This includes, among other objectives, achieving social balance. This is necessary to meet the needs of Venezuelan society. There is a requirement both to transform the material and social conditions of the majority of the population, historically separated and detached from equitable access to wealth and welfare, and to build a new citizenship status, based on full recognition and exercise of the rights guaranteed to citizens, as human beings and social subjects with autonomy in all spheres of social life. This is a different rationale from what has occurred in the past and continues to exist in many places today, that is, respect for and protection of and defence of the quality of life for all inhabitants of Venezuela. It is necessary to deconstruct the notions of universal social rights, equity of access to resources, means of support and the material and welfare conditions, since the practice is far from the theory. On the other hand, the humanistic character on which these ideas are based requires a thorough review of the interpretation of these concepts in today's society.

Since education is a right established in the Constitution of the Bolivarian Republic of Venezuela, there is in the Plan for Economic Development and Social Welfare (PDESN) a guarantee for access, permanence and continuity in response to social needs so as to ensure the conditions of universality with equity. This establishes that the quality of education is conceived from two perspectives: formal quality and political quality. The former concerns the technical and scientific capacity, the development of learning capabilities with regard to content and the use of methods of academic relevance. The latter, on the other hand, refers to the development of citizens, empowering them with the values of solidarity, participatory and protagonist democracy that lead to the construction of citizenship: the ability to be the subject of individual and collective social action, to organise themselves through associations and cooperatives, to cultivate cultural identity, with universal and critical significance to obtain and practise their rights.

As part of these objectives, it is proposed: to guarantee access and permanence in the educational system; to extend the coverage of enrolment levels and modalities; to articulate education with systems for producing goods and services; to develop its

[3] Luis Beltrán Prieto Figueroa (1902–1993), the educator and Venezuelan politician who struggled hard for education for the masses.

infrastructure and educational resources in different levels and modalities; to achieve equity in higher education; to provide for the comprehensive care of children and youth not included in the education system; and to eradicate illiteracy.

The foundations laid down in the Constitution and outlined in the Plan of Economic and Social Development of the Nation 2001–2007 demand a re-conceptualisation of education in general. This is not just from within but also from the vision and understanding of it from a global perspective and the influence it exerts on international organisations of which Venezuela is a member. Since 1959, developed countries have been implementing standardised tests in order to compare the quality levels of educational systems based on standards that aim to measure the performance levels in mathematics and science to varying degrees. Among these tests are the TIMSS (Trends in International Mathematics and Science Study) developed by the International Association for the Evaluation of Educational Achievement (IEA), the IALS (International Adult Literacy Survey) developed by the Organisation for Economic Cooperation and Development (OECD) (1994) and the Latin American Laboratory for Assessment Quality of Education (LLECE), founded in 1994 and coordinated by the Regional Bureau for Education in Latin America and the Caribbean, UNESCO. The latter organisation has proposed an 'ideal profile' school that would allow greater achievement levels of students. All of the studies mentioned above and other new initiatives point to measuring the achievements of students associated with their participation in society. As standards driven assessments, they are directly linked to a single component of the curriculum, the official, leaving out many influential elements in education.

The Constitution framed the nation in mainstream humanistic philosophy and it is necessary to recover the true meaning of many of the concepts found in documents generated by summits, conventions and international organisations. In practice, each of these concepts has an accepted meaning *per se*, removed from the reality in which they may be applied. Globalisation has permeated everything and, as defined by Castro (1999), it is the unity of all inequalities. In this sense, the quality standards of education established by international organisations do not necessarily have to match the new reality as described in the legal framework in Venezuela. It is imperative to review what is meant by quality of education in the international framework and the current order; it is also necessary to define whether these measures correspond to the reality of the Venezuelan people.

Venezuela is a member of several international institutions intended to improve the minimum conditions for education in the participating nations. Among these organisations is UNESCO, which in March 2001 convened in Cochabamba (Bolivia) the 7th Meeting of the Intergovernmental Regional Committee of the Major Project Ten-Year Education in Latin America and the Caribbean (PROMEDLAC VII). Present at that meeting were other members from different regions, UN agencies, inter-governmental and non-governmental organisations and representatives of various institutions and foundations (Declaration of Cochabamba, UNESCO 2001a, b). In that meeting participants discussed the results of the evaluation of 20 years of the Major Project of Education in Latin America and the Caribbean as well as the possible political scenarios, social, economic and cultural environments in which education would be developed in the region over the next fifteen years. However, it was not possible for such a meeting to foresee the political, economic, social and cultural development of the Bolivarian Republic of Venezuela for the period 2001–2015, and the qualitative or quantitative advancements in each of the scenarios studied for the country. This meeting led to a series of recommendations on education policy at the beginning of the twenty-first century, demarcated into nine sections:

1. The new meaning of education in a globalised and constantly changing world;
2. Learning and attention to diversity in quality: priorities of education policies;
3. Strengthening and redefinition of the role of teachers;
4. Service management processes of learning and participation;
5. Expansion and diversification of learning opportunities throughout life;
6. Media and technology for the transformation of education;
7. Funding for quality learning for all;
8. Information systems for the improvement of educational policies and practices;
9. International cooperation.

The changes started in the CRBV and the missions and projects that have been created by the Venezuelan government can be looked at in the context of these nine sections. Besides, they have been reinforced in the LOE approved in 2009. Simultaneously these issues are also part of the agenda of several government ministries and secretariats. A new meaning to education has been given that encompasses principles of inclusion, multiculturalism, multiethnicism, and plurilingualism with a strong qualitative focus based on humanistic principles. The education or teacher training programmes have been increased significantly at universities and the funding for teachers and students at all levels has been impressive. An example of policies is the LOE itself plus the projects created. An example that combines international cooperation, education policies, teacher training and the humanistic perspective is the creation of the Iberoamerican Sports University initiated in 2006 where students from Venezuela and low income developing countries come to study under scholarship programmes.

The Bolivarian Education System

Start the social building by its basements!
Not by its roof.... Children are the stones

Simón Rodríguez (Rumazo 2008)

'Education for All' is the main goal of the Bolivarian Education System, in contrast to the neo-colonisation, acculturation, alienation, elites and exclusion that characterised previous educational arrangements. The Bolivarian education system generates a new emphasis on State-Family-Society, where the school is a space that focuses the actions and organisation of state power, promoting participation, making necessary changes in institutions and culture, strengthening an endogenous development model and sovereignty, correcting imbalances and achieving environmental sustainability and a decent quality of life. All these factors are united in this new education paradigm which is central to people as social, responsive and active participants in the transformation of Venezuelan society. Education is conceived as a human continuum in which services, teaching and learning processes are a complex unit having corresponding levels and modalities. Education covers the stages of development of each individual looking at his/her physical, biological, psychological, cultural, social, environmental and historical characteristics. The Bolivarian education system seeks to eliminate the social debt generated by the system of exclusion in order to create a social equilibrium model. Ongoing programmes ensure access, retention and continuing education for everyone in the education system. The State, together with the family and society, universalises the right to education as a human right and a fundamental social duty (Rojas 2005). Basic education according to the new Law of Education (2009) refers to three stages of education: initial education (0–6 years old), primary school (6–12 years old: 6 grades) and high school (13–18 years old: 5 or 6 years). In the following paragraphs we present a brief summary of the different programmes.

Simoncito (Little Simon Project) Project has been developed by the Ministry of Education and Sports of Venezuela (starting in early 2004 and still continuing) with the aim of providing the nation's children with the early education and care they need from pregnancy up to 6 years, with the participation of their families and their communities. This initial education is divided into two sub-levels: maternal (0–3) and pre-school (4–6 years old).

The Bolivarian Schools Project is a Venezuelan state policy that identifies education as a human continuum. It comprises initial education (Simoncito) comprising the first 6 years/grades of school. The Bolivarian High School is an educational institution that prepares students in their infancy and adolescence for endogenous development and sovereignty through a new conception of school that emphasises Bolivarian identity and citizenship. The Robinsoniana Technical Schools conceive education and work as fundamental processes for technical and personal development, respecting the dignity of students and preparing them for the democratic exercise of popular will and the construction of a fair, peaceful and loving society. There is a strong tendency to look at areas related to agriculture and in recent years they have also included such areas as nursing, accounting and mechanics.

Ribas Mission is an educational programme promoted by the Ministry of Education of Venezuela in order to include all those who have been unable to complete high school. It uses the system of "teleclass (recorded TV class)" which provides audio-visual aided instruction led by a facilitator. Mission Robinson "I can" was aimed at eliminating illiteracy among young people and adults across the country. The second phase of the Mission (Robinson II) has as its objective the completion of the sixth grade at school by all participants, as well as the consolidation of literacy.

Sucre Mission aims at decentralising higher education. It aims at providing university education to all regions and localities in the country so that the great mass of high school graduates who, so far have been excluded, may benefit. They will be incorporated or enabled to continue their studies in higher education.

The Strategic Plan for Information Technology and Communication in the National Education Sector (PETISCEN) aims to increase the quality, democratisation and national relevance of the educational process and the effectiveness and efficiency of administration and management of the Ministry of Popular Power for Education (the name was changed in 2008).

The idea of a university for all suffered in the last decades of the last century, as the State reduced its responsibilities for education. From 1989 to 1998, investment in higher education in Venezuela showed a downward trend affected by the propensity to make budget cuts in all sectors of the social field. Particularly, this was a strategy for building sustainability of the proposed privatisation of higher education. This resulted in a large social debt accumulated as university enrolment decreased, resulting in the exclusion of students from the poorest sectors. Admission to higher education sectors favoured those with higher incomes, the residents of large cities and students from private schools (Fuenmayor and Vidal 2000). Along with this phenomenon, the management of higher education changed dramatically in favour of private management of education. However, since 1999 the trend in the decline of the public budget dedicated to education has been reversed, increasing from 3.2% of GDP in 1998 to 4.6% in 2002 and 7% in 2005, in addition to the budget allocated to the Sucre Mission and the creation of new universities. This percentage is significant as compared to the whole of Latin America, which spends on average 4.4% of GDP on education, though it has a long way to go to achieve the levels of investment in the European Union or the United States of America.

Conclusion

An ignorant people is the blind instrument of its own destruction.

Simon Bolivar (Prieto 2002)

The Bolivarian Education System is a way of achieving the United Nation's millennium goals. As referred to earlier in this chapter, the CRBV establishes the citizens' right to education and the importance of public schools in social integration. Education is both compulsory and free. The CRBV acknowledges the rights of groups and marginalised minorities and proposes ways of providing citizens with access to empowerment. It also proposes the universalisation of rights and social balance in providing the same rights to all (MECD 2004). Education is understood to be a human right, without any sort of discrimination and the government is providing the means for achieving this through more schools, high schools and universities, staff, programmes and resources, so making it possible to speak of equity as an ethical and political principle.

Quality of education is important but it is far more important to pay the huge social debt in the country. Venezuela has a history of inequality, inheriting a huge gap that separates those who could and those who could not have access to education.

That is why it is so relevant that the government has created programmes/missions to provide education for all, meaning an education of quality and equity. In order to achieve this objective, there is a shared responsibility between the State, Family and Society as the foundation of a participatory and protagonist democracy. In this way solidarity and equity are attainable with the purpose to achieve social balance, promoting a state of justice and giving a social context to this democracy in the context of humanism.

Bolivarian education is the centre of the trilogy State–Society–Territory that should promote cultural and institutional change in order to consolidate the model of endogenous development through social production, correcting imbalances and inequalities and creating a sustainable environmental society. The previous plans of the nation (before 1999) did not achieve the benefits for most of the citizens because there was a high percentage of illiteracy, exclusion from school and poor pre-school education. The Bolivarian education was intended, and it is already happening, to solve the educational social debt problem through the different missions now in place. This is changing the existing society and is creating a more balanced development of the citizens. Before it was hard to speak about or understand the concept of sustainable society or environmental education, nowadays it is changing and there are more discussions and programmes in place with strong support in legal documents and policies, but it takes a while for people to be empowered and to change the way they used to consider or refer to the environment. The environment is not theirs; it is ours and is the only place we have in which to live.

Under the Constitution of the Bolivarian Republic of Venezuela, approved in 1999 by the majority of the people, the PEN-based project developed by the teachers and the Plan of Economic and Social Development of the Nation 2001–2007 gave direction to the strategic work plan which includes restructuring the SEO and the school as a community centre (Community Schools have these dimensions: economic, cultural, community, educational, social and political). It guarantees free, compulsory and quality education (for the first time children and young people do not pay for their registration at public schools). Similarly, management teams have been set up to change the old centralised management structure. It is proposed that all decisions should be made after being analysed, reflected on and evaluated by the respective responsible team, what is called horizontal management, also described as participatory democracy. Venezuela introduced a new model of governance and a new model of teacher status which is the fundamental commitment to social transformation from the state apparatus called education. The nation has experienced strong, complex, violent and difficult historical moments but the greatest challenge is to change state structures to facilitate social transformation and the welfare of all Venezuelans. This means a rebirth of the Republic, i.e., a revolutionary process that requires all citizens to review their conceptions and actions in the framework of the constitutional principles. It has been difficult, but while keeping to the principles of solidarity, respect, brotherhood, love, cooperation and transparency, some mistakes have been made, but these have been recognised and problems are being overcome.

In the last 10 years, important progress in education has been made, such as: the privatisation scheme was abandoned and the public nature of education re-established;

the fight towards exclusion and the promotion of inclusion of all; investment is increasing to meet the construction and repair of school buildings; increased attention is being given to early childhood education; greater access to technology; over a million new literates; improvement of special education programmes and technical schools; priority attention to intercultural, bilingual education; and the extension of higher education by creating university campuses of already established universities in various municipalities to allow students to study in their own places and to avoid the migration to the central states of the country in order to study. In addition there has been significant progress in the use of technology, school enrolment growth, and increased incorporation of adult education through educational missions. Finally the long awaited Organic Law of Education under the guidance of the Bolivarian Constitution has approved the importance of environmental education and sustainable development. However, there are many more things to be achieved and there is a need to protect and nourish what has been achieved because, as Gramsci's (Biondi 1977) famous phase indicates, "…the old model has not completely finished dying and the new one has not been completely born yet".

References

Asamblea Nacional Constituyente. 2006. *Constitución de la República Bolivariana de Venezuela*. Caracas: Ministerio de Comunicación e Información.

Biondi, M. 1977. *Guide Bibliografica a Gramsci*. Cesena: Librería Adamo Bettini.

Castro, G. 1999. *El Asalto del Plural*. Caracas: Faces UCV. Tropykos.

Embassy of the Bolivarian Republic of Venezuela to the UK and Ireland. 2009. Venezuela's New Education Law: Myth and Reality. Available at http://www.venezlon.co.uk/pdf/fs_Ley_educacion.pdf. Accessed 1 May 2009.

Figueroa, L B P. 1959. *El Humanismos Democrático de la Educación. IESALC-UNESCO*. Caracas: Fondo Editorial IPASME.

Figueroa, L B P. 2002. *El Magisterio Americano de Bolívar*. Caracas: Universidad Pedagógica Experimental Libertador.

Fuenmayor, L, and Y Vidal. 2000. La admisión estudiantil a las universidades públicas venezolanas: aparición de iniquidades. *Pedagogía* XXI(62): 273–291.

Fundación Casa de Bello. 1992. *Simón Rodríguez*. Caracas: Ediciones Casa de Bello.

López de D'Amico, R, and A González. 2006. Venezuela: A permanent question to education. In *Education reform in societies in transition*, ed. J Earnest and D Treagust. Rotterdam: Sense Publishers.

Loreto, M. 2004. *Registro histórico de la educación en Aragua, 1999–2003 (Resumen). Unpublished document*. Maracay: Zona Educativa del Estado Aragua.

Ministerio de Educación Superior (MES). 2002. *Universidad Bolivariana de Venezuela. Curso básico. Educación Superior para el Desarrollo Endógeno*. Misión Sucre.

Ministerio de Educación, Cultura y Deportes. 1999. *Organic Law of Education [LOE]*. Caracas: Ministerio de Educación, Cultura y Deportes.

Ministerio de Educación, Cultura y Deportes (MECD). 2002a. *Constitución de la República Bolivariana de Venezuela*. Caracas: Ministerio de Comunicación e Información.

Ministerio de Educación, Cultura y Deportes (MECD). 2002b. *Orientaciones Para el Proceso de Discusión Curricular. Años Escolar 2002–2003*. Caracas: Ministerio de Educación, Cultura y Deportes.

PDESN. 2001. *Líneas Generales del Plan de Desarrollo Económico y Social de la Nación 2001–2007*. Caracas: Ministerio del Poder Popular de Planificación y Finanzas.

República Bolivariana de Venezuela. 2004a. *La Educación Bolivariana. Políticas, Programas y Acciones. "Cumpliendo las metas del milenio"*. Caracas: Ministerio de Educación y Deportes.

República Bolivariana de Venezuela. 2004b. *La Educación Bolivariana. Políticas, Programas y Acciones*. Caracas: Ministerio de Educación y Deportes.

Rojas, A. 2005. *La Educación Bolivariana*. Caracas: Ministerio de Comunicación e Información.

Rumazo, G A. 2008. *Simón Rodríguez. Maestro de América*. Caracas: Biblioteca Ayacucho.

UNESCO. 2001a. *Declaración de Cochabamba y Recomendaciones Sobre Políticas Educativas al Inicio del Siglo XXI*. Cochabamba: UNESCO.

UNESCO. 2001b. *Educación para Todos en las Américas. Marco de Acción Regional*. Dominican Republic; UNESCO.

Part II
Schooling for Sustainable Development in Brazil

Chapter 7
The Environment as a Cross-Curricular Theme in the Brazilian "Parâmetros Curriculares Nacionais" (PCNs, National Curriculum Parameters)

Maria Lucia de Amorim Soares and Leandro Petarnella

Brief History of Brazilian Educational Trends

This brief overview of the main Brazilian educational trends places the educational reform of the 1990s in the historical trajectory of Brazilian education. To discuss trends in educational philosophy of education that influenced and still influence the pedagogical practices of educators, Saviani (2000) identifies five major trends that characterise the history of Brazilian education:

1. *The Humanist Tradition Project* is based upon the educational work of the Jesuits, the key planners of Brazilian education and the reforms of the Government of Pombal in 1759. Despite the expulsion of the Jesuits from Brazil, the Jesuit educational assumptions persisted in the traditional lay pedagogy. In the dynamics at work in traditional pedagogy the teacher has unquestioned authority derived from the domain of knowledge.
2. *A Humanist Modern Project* found expression in the early twentieth century and inspired the New School movement. It gained notoriety with the release of the *Manifesto dos Pioneiros* (Manifesto of the Pioneers) in 1932. Until the 1930s, the forces of the old republic were dominant in Brazil with power centred on agrarian oligarchies and education was the privilege of a few selected people. The lower classes were generally illiterate with only a small number of children having access to primary schools. Those schools that did exist were outdated in relation to the economic needs generated by the advance of capitalism and they became principal contributors to the backwardness of the country. From this perspective, as in a model of capitalist society, there is a typically capitalist school

M.L. de Amorim Soares (✉)
University of Sorocaba, (UNISO/SP) Sorocaba, Brazil
e-mail: maria.soares@prof.uniso.br

L. Petarnella
University Nove de Julho, Sao Paulo, Brazil

M.L. de Amorim Soares and L. Petarnella (eds.), *Schooling for Sustainable Development in South America*, Schooling for Sustainable Development 2,
DOI 10.1007/978-94-007-1754-1_7, © Springer Science+Business Media B.V. 2011

facilitating the achievement of this model. Poverty and extreme social inequality are the result of a slow and imperfect realisation of that vision of a society without the help of appropriate school and social reform.

The *Manifesto dos Pioneiros* (Manifesto of the Pioneers), written and embraced by many intellectuals in 1932, called for a reduction in illiteracy and ignorance. But it is only in the 1960s that this approach emerged with the presentation of guidelines different from those for traditional teaching, especially with regard to new roles for both the student (the subject of an active and more dynamic learning process) and the teacher (no longer the sole source of knowledge). Such propositions came to be incorporated into the practice of most teachers, especially in the public schools. Some schools lacked the physical infrastructure and appropriately trained teachers necessary to introduce pedagogical changes in the direction advocated by the so-called pioneers of education.

3. *The Technical Project* appeared after the military coup of 1964 when the military assumed power against the 'communist threat' that was said to exist in Brazil. The centre of the country was opened up for economic development by multinational enterprises as a result of agreements between the Ministry of Education and Culture and American agencies. In line with the ideas underlying such agreements, the enactment of Law 5672/71 was intended to realise the project of professionalisation of teaching, the training of manpower, until then predominantly rural, to work in multinational companies and to formalise educational arrangements inspired by behaviourism yielding a technical approach. The principles underpinning this technical approach were rationality, efficiency and productivity. To put these principles into practice, programmes and actions were orchestrated and initiated by experts, thinkers and teachers of educational policy. In practice, the Brazilian school system was given the chief task of modernising elite education to prepare students more effectively for leadership in a more complex society. It was intended to provide executives from the ruling classes with a modern mind, a sound general knowledge and intellectual skills that would enable them to perform the task of imposing new forms of production and new working relationships on favourable terms for foreign and domestic companies.

   As a complementary function, it was expected that the school would qualify the manpower, within the limits imposed by the size of the nation's industrial development and labour supply and the restrictions linked to increasing technological sophistication. The solution to the problem of having sufficient trained manpower was found in the creation of vocational high schools. However, these were inefficient and inappropriate to the needs and possibilities of the working class. Pragmatic measures were introduced, like the creation of a parallel system of vocational training, organised and maintained by businesses according to their interests and needs
4. *Critical Project-Reproducible*, This was based on the theories of Bourdieu and Passeron (1970) (with an emphasis on the symbolic violence of the education system), Althusser (1983) (with reference to the school as the ideological apparatus of the state) and Baudelot and Establet (1978) (focusing on the theory of the two-tier school). From these theories, the school was now being studied from the

perspective of its determinants and perceived as an instrument of social domination. Thus, education was understood in the primary instance as part of the proper functioning of the capitalist system.

5. *The Historical-Critical Pedagogy* highlighted the paralysis of Critical Project-Reproducible. It pointed to the fact that the previous project did not advance in a critical sense, i.e., it did not formulate alternative solutions to the problems that had been identified. Critical pedagogy made history at the end of the 1970s because these views on education stood for pedagogical and social practices that were committed to the transformation of society. The dialectical conception of such a pedagogical stance, i.e., the perception of the contradictions between school and society, generates and develops this historical-critical-pedagogy.

However, given Brazil's engagement with the broader international educational movements resulting from its increasingly important participation in the global competitive economy, a return to a liberal conception of education resurfaced with a new pattern in Brazil. There was a need to defend such pillars as interdisciplinarity, contextualisation and skills development against being co-opted by powerful interests related to late capitalism. In this scenario, the Brazilian educational reforms of the 1990s are in line with the demands of the large industrial organisations with regard to improvements in the quantity and quality of education of the workers and the scientific and technological standards of professional training.

The Brazilian Educational Reforms of the 1990s

With the introduction of a new national Constitution in Brazil in 1988 the education system was restructured. Law 9.394 of 1996 established the "Lei de Diretrizes e Bases da Educação Nacional" (LDB, Law of Directives and Bases of National Education) that resulted in the following important changes:

- Democratic management of the public education system and independent management of the educational units;
- Compulsory and free elementary and primary education;
- A minimum attendance of 800 hours per 200 days at elementary school;
- A common curriculum core for elementary schools and high schools complemented by a curriculum component that took into account local circumstances;
- College teachers education (graduation), or PhD specialisation;
- Budgetary reserve of 18% of gross domestic revenue coming from the Union (the management organ that was part of the Federal Government's responsibility) and 25% coming from the States and Municipalities with the purpose of promoting the development of public education;
- Creating a National Plan for Education.

In the 1990s, the LDB established the "Parâmetros Curriculares Nacionais" (PCNs, National Curriculum Parameters) for each subject to be taught compulsorily in all grades of elementary schools and high schools, and in all kind of schools – public schools, private schools, confessional schools and community schools. The PCNs are a group of proposals relating to the organisation and development of the school curriculum to be followed by school managers and teachers. The PCNs are flexible, aimed at the attainment of a reasonable level of achievement for the national education system, taking into account the differences in the social and cultural realities of each school and its region.

Linked to the PCNs, Cross-Curricular Themes were established according to the social diagnosis of the time, with the intention of promoting discussions about topics that are shared by various school subjects as well as stimulating reflection about national realities. The overall goal was citizenship formation and the themes included:

- Ethics (meaning shared common values)
- Environment (care of the places where we live);
- Cultural plurality (a guarantee of the right of the individual to be different);
- Health (the quest for universal well-being: mental, physical, spiritual, according to a definition provided by the World Health Organisation;
- Sexual guidance (to counter the lack of attention to sexual education in schools that was seen as contributing to premature pregnancies, the numbers of which were causing alarm).

The National Curriculum can be interpreted as a group of reform proposals created to serve as the basis for the elaboration and revision of the subject content policies of the states and municipalities. They serve as guidance as to the way the states and municipalities can change their local and regional educational systems, proposing educational and methodological principles to be observed by public educational institutions. In spite of the many criticisms of the weaknesses in the document – especially where reference is made to a positivistic conception of knowledge (Macedo 1999) – the inclusion of the environmental theme in this process, taking account of its range, has been an important achievement with regard to environmental education. It is important to point out that this inclusion took place in the very first version of the PCNs. This was not the case for other cross-curricular themes, such as the one referring to 'cultural plurality'. The main purposes of the PCNs – structured on the areas of Portuguese language, mathematics, natural sciences, history, geography, arts, physical education and foreign language – have been difficult to implement. It had been assumed that some students would be able to achieve at least an understanding of citizenship as social and political participation gained from the following:

(a) taking sides, in a critical and responsible way over different social situations, through dialogue;
(b) knowing the fundamental characteristics of Brazil and their different dimensions;
(c) understanding and valuing the plurality of the Brazilian cultural patrimony;
(d) realising themselves as part of the environment and as individuals with responsibilities for making environmental changes that can help its improvement;

(e) developing their self-confidence and self-criticism; through the adoption of healthy habits, a consciousness and an appreciation of themselves;
(f) using different forms of expression (oral, mathematical, body language, etc.) as ways to produce and express their ideas;
(g) being able to use different sources of information and the technological resources that are necessary to achieve and generate knowledge;
(h) questioning reality, identifying problems and finding solutions for them, through logical reasoning, creativity, intuition and critical analysis skills (Brazil 2002: 107).

The National Curriculum document does not specify ways for the schools to achieve these aims with their students. It does not provide schools with advice about asking the "Secretarias de Educação" (Department of Education) and other responsible organisations to make available the necessary measures to guarantee their fulfillment. Carvalho (2003: 191) affirms that the document offers,

> An important recognition of importance. Recognition that the classical disciplines as representative areas of scientifically accepted knowledge by themselves are no longer sufficient to meet the challenges necessary to achieve this. The "solution-antidote" to get around these "problems" is the presentation of a set of themes called cross-curricular, justified because they encompass the whole range of disciplines.

The concept of cross-curricularity that the National Curriculum adopted promotes a basic difference between the interdisciplinary and the cross-curricular proposals. Interdisciplinarity rejects entirely disciplinarity, assuming that interdependence of the different kinds of knowledge is essential to the utilisation of other disciplines, either in the joint action of two or more disciplines, or the option of working with a stimulus theme that can be analysed through its interconnection with different disciplines. Cross-curricularity not only admits disciplinarity sometimes, but also supports the practice where some questions should be answered by each knowledge branch or a particular knowledge branch (Cordiolli 1999). The main characteristic of cross-curricularity is the linking of these questions/subjects with everyday experience at the exact moment that that knowledge is being built inside the classroom providing immediate answers to the questions that are asked (Carvalho 2003). Cross-curricularity is based on the assumption that the subjects in different branches should not create an isolated knowledge but should value a full understanding of each theme. Further, it is assumed that, because of a theme's dimension, range and complexity, packed with conflicting values and interests (Sancho 1998), a space is opened up for establishing a new dialogue inside the classroom. It also facilitates the valorisation of the knowledge achieved out of school and its possible connection with the formal educational programme.

However, despite the consensus achieved over the acceptance of cross-curricular themes in the PCNs (National Curriculum Parameters), the nucleus of the educational programme continued to be based on the classical disciplines. It was difficult to replace the traditional approaches, even when a strong case had been made to adopt a broader basis for the analysis of contemporary issues, bringing studies in schools closer to the realities of the third millennium. As a result, the cross-curricular themes, having an undeniable importance for the formation of critical beings and active citizens, will only be included as an option in the educational programme if there is a will to do it. This reinforces the maintenance of the *status quo*.

The Environment as a Cross-Curricular Theme

Despite the many ways of defining environmental education, it can be understood as that part of education directed at environmental questions, or as,

> One dimension given to the content and practice of education, geared to solving practical problems of the environment through interdisciplinary approaches, and the active participation of each person as a responsible individual in the community (Dias 1992: 31).

So, it is a kind of education where specific processes with specific aims, purposes, strategies and contents take place, characteristics that show it as an active and lifelong education able to promote a reaction to,

> ...The changes that take place in a rapidly changing world. Such education should prepare individuals, through the understanding of the main problems of the contemporary world, providing them with technical knowledge and necessary skills to play a productive role with a view to improving the quality of life and protecting the environment, paying due attention to ethical values (*ibid.*: 68).

Environmental education appears as a new paradigm of education, bringing it to the twenty-first century and offering an education more adaptable to the new demands of society and the environment. It challenges educators to bring together simultaneously two different voices,

> The first, marked by a voice, reforming, revolutionary, transforming, using it to justify keywords in the contemporary imagination, like complexity, community participation, quality of life, sustainability, and the second voice that is grounded in a conservative discourse, backward and somewhat nostalgic, where there are other keywords such as conservation, heritage and tradition (Sanchez 2003: 49).

It is important to point out that the polyphonic, ambiguous and paradoxical discourse generated about environmental education leads to the questioning of its epistemological limits (Brugger 1994), where education takes on a technical vision that does not allow a clear distinction between *education* and *training* with regard to environmental questions. But it is environmental education, as a common project bringing together people from various social classes, which promotes a future project for humankind based on technical and scientific justifications. As such, it advocates new habits, attitudes and human behaviour with regard to environmental issues.

As an example, we can refer to an extract from the *Declaração da Intergovernamental de Tbilisi sobre a Educação Ambiental* (Tbilisi Intergovernmental Declaration on Environmental Education), an official document published by UNESCO (1977) and signed by over 140 countries during a famous international meeting which took place in 1977. This is considered by many to be one of the main theoretical and conceptual landmarks in environmental education and the relevant extract reads,

> Through the use of advances in science and technology, education must play a critical role in order to create awareness and better understanding of the issues affecting the environment. This education is to foster the development of positive behaviours of conduct with respect to the environment and the use of its resources for the nations (quoted by Dias 1992: 68).

It was through the creation of the "Programa Internacional de Educação Ambiental" (PIEA: International Programme for Environmental Education), approved during the United Nations Conference for the Human Environment in Stockholm (1972), that, for the first time, the importance of an educational measure for addressing environmental questions was officially recognised. Since then, a series of international conferences, including those in Belgrade (1975), Tbilisi (1977), Moscow (1987) and Rio de Janeiro (1992), have confirmed the importance of the concept, defending formal education as one of the essential axes for environmental education and insisting on its inclusion in the educational systems all over the world.

The way environmental education has been debated and stimulated by the PCNs represents a way of making this inclusion come true, recognising that schools play an important role in the formation of active, responsible citizens, promoting essential values like ethics, fraternity and respect for life. On account of its holistic approach, it is managed in a more noticeable way in the PCNs through the proposition of the environment as a cross-curricular theme. The legitimisation of the implicit cross-curricularity of the environmental theme is an essential step to not allowing environmental education in the schools to be thought of and reduced to just one more discipline derived from the biological sciences. Environmental problems are related not only to the protection of life but also to the quality of life. Carvalho (2003: 94) suggests,

> Thus the unequal distribution of income and widespread social injustice arising from the development model adopted, the technology race, individualism and the growing challenges of cultural diversity that characterises today's globalised world, are also regarded as environmental.

In choosing the environment as a cross-curricular theme there were several basic motives, including: social urgency; national scale questions; the teaching/learning opportunities in elementary schools; the need to facilitate an understanding of reality; and the need to promote active participation, especially with regard to having the ability to face and take sides on questions that interfere with life in society. Considering these motives, it is not so difficult to understand why this inclusion happened. It reflected an awareness of the seriousness of contemporary ecological questions and the urgent need to build a consciousness that is sensitive to guaranteeing the survival of humankind. In a way, this proposition reflects the need to develop environmental education practices that occur in a spontaneous and conscious way.

The presentation of the cross-curricular theme 'Environment' in the PCNs has two well-defined parts. First, there is the history of those significant conferences and world meetings concerned with the environment where environmental education had been highlighted as a key to the formation of new attitudes relevant to environmental matters. Secondly, environmental education directed at elementary school teaching is discussed with regard to its contents, assessment criteria, educational guidance and teaching practices. Environmental education is defended as a revolutionary proposition that, when well used, can lead to "personal behaviour changes and to actions and values of citizenship that can cause strong social consequences" (Brazil, PCNs, vol. 9: 27), as demanded by the National Constitution and declared as a priority in all jurisdictions of government.

The PCNs recognise that environmental education is still far from being "an easily developed and accepted activity" (*idem.*). Without changes in political will, in inhuman and globalised capitalism, in uncontrolled consumption and in selfish individualism, environmental education alone cannot act like magic. At the same time, civil society's role is not sufficiently clear in the initiatives of a variety of associations, NGOS, action groups and popular institutions, but there is an assertion that Brazil is one of the countries that develops many "original initiatives" (*idem.*: 41–49), including,

1. Respects and takes care of living beings.
2. Improves the quality of human life.
3. Preserves the vitality and diversity of the planet.
4. Minimises the exhaustion of resources.
5. Respects the planet's limitations.
6. Changes negative actions and practices relating to the environment.
7. Allows communities to take care of the environment.
8. Creates a national framework for integrating development and conservation.
9. Facilitates the construction of a global society and biodiversity.

In the second part that focuses on contents and educational guidance there is a list of the aims to be reached at the elementary school level (*idem.*: 53–54):

1. Integrated knowledge and understanding about basic notions of the environment.
2. Adoption of sustainable attitudes at home and at school compatible with that understanding.
3. Observation and critical analysis of environmental facts and situations relevant to the topic.
4. Realisation of cause and effect phenomena in nature, important to understanding the environment and its different ecosystems.
5. Control of preservation procedures and the exploitation of natural resources.
6. Realisation and valorisation of social, cultural and natural diversity and personal identification as an integral part of the environment.

The contents related to the environmental theme were divided into three cycles:

- nature cycles, approaching the group of interrelations and flows existing in nature through systematic observation;
- society and environment, directed at the study of the interrelations between human groups and the activities they develop in a determined space;
- and environmental protection and preservation, from a more practical point of view directed at the analysis of the incentives to adopt practices that respect the environment and avoid waste.

The inclusion of the environmental theme as a cross-curricular topic has another positive aspect in addition to promoting formal environmental education. It has the potential to draw to the attention of the school as a whole the importance of the theme and the responsibility of everyone for it. The decisive elements of social and environmental values learning are to be found in everyone living together in the school community and in the classrooms (*idem.*: 57, 75). However, while offering some clues, the PCNs do not give concrete directions as to how such changes can be introduced. Each school must rely on its own creativity in its search for better ways of responding to the demands of its own reality.

Eckersley (1992) has broadened the understanding that environmental problems came from a cultural crisis, or a crisis in the structure of social values. For theorists of environmentalism it was necessary to integrate the concerns of the environmental movement with other social movements, such as peace, feminism, and development aid to the Third World, in order to find ways to overcome the consumer values systems of domination and the logic of capital accumulation.

In Brazil, amid bouts of military rule following the 1964 coup, the history of environmentalism occurred differently. Especially after the Stockholm Conference (1972), issues concerning the environment began to appear more frequently in the media in the South and Southeast, the more developed parts of the country, with demonstrations against pollution and environmental damage. Despite these demonstrations, as indicated by Viola (1997), during the 1970s, the environmental movement made little impact on Brazilian public opinion. However, since 1981, when the country ceased to be the champion of economic growth, the impact on society and on the production of ideas was great, marking the growth of environmental awareness and environmentalism. At that time, it could be seen that the development model adopted was one that polluted and degraded natural resources and produced significant social inequalities. As a result, in the second half of the 1980s Brazilian environmentalism was no longer restricted to small groups in civil society and governmental bodies; it became multi-sectored permeating other social movements, non-governmental organisations (NGOs), universities, the media and government agencies that are not specifically engaged in environmental and business activities.

The year 1988 marked a milestone in Brazilian environmental policy as the Federal Constitution (Brazil 1997) devoted a chapter to the environment, which includes,

> Article 225. Everyone is entitled to an ecologically balanced environment and the common use and essential to a healthy quality of life, imposing upon the State and society the duty to defend and preserve it for future generations.

With specific reference to environmental education, the Government issued instructions to promote environmental education at all levels of education and public awareness for the preservation of the environment. Finally, in 1992, 20 years after the Stockholm Conference, the city of Rio de Janeiro hosted the First United Nations Conference on Environment and Development (Rio–92) and this is considered to be a milestone in international efforts directed towards sustainability. Its main objective was the elaboration of strategies seeking to halt and reverse the effects of environmental degradation. At the conference there were five important documents:

1. The Rio Declaration, also known as the Earth Charter. Statement of principles in which rights are defended and government and citizens are rendered accountable in relation to the environment.
2. Agenda 21. Roadmap for action that sets out goals to be achieved in the twenty-first century, aiming to direct the transformation of society toward sustainability.
3. Declaration of the Forests. Statement which outlines principles for sustainable forest conservation. Not formulated at the conference due to divergent opinions.
4. Convention on Climate Change. Convention outlining measures to reduce greenhouse gas emissions from burning fossil fuels and, consequently, to reduce global warming.
5. Convention on Biodiversity. Convention which defines the need for *in situ* conservation of biodiversity, providing for the rights of countries with significant biodiversity.

Rio–92 provided a forum for unofficial discussions between representatives of non-governmental organisations and civil society that resulted in the Treaty on Environmental Education for Sustainable Societies and Global Responsibility. The treaty emphasised that environmental education is not neutral, but ideological, is a public act based on values for social transformation.

Lessons in Sustainability and the Projeto Escola do Amanhã (Schools for Tomorrow Project)

Studying waste disposal and engaging in field research to assess the effectiveness of rubbish collection are among the innovations in environmental education promoted by schools in the Schools for Tomorrrow Project. Work with students has extended beyond the separation of recyclable waste, as these examples of activities undertaken in schools in the city of São Paulo, São Paulo State, illustrate:

1. Conscious consumer: students make PET bottle piggy banks, saving money and in the middle of the year, discussing with their family how to use it, respecting the idea of conscious consumption (Magister College on the south side of the town).
2. Junk: lessons in the science laboratory include the dismantling of equipment that has become junk and the separation of what can be recycled (Pentagon College, which has sites in São Paulo and Greater São Paulo-Metropolitan).
3. Statistics: statistics on the content of landfill disposal are used to assess what has been generated in homes, raising student and staff awareness (Chestnut Trees School in Greater São Paulo).
4. Solar heaters: students participate in a programme that installs low-cost solar heaters in the homes of low income families (Santa Maria College in the south of the city).

5. Committees of students: a students and faculty committee works on themes concerned with garbage, water, energy and material and formulates proposals to implement a car-pooling scheme among students' parents (St. Louis College in the area of Paulista Avenue, São Paulo).
6. Campaigns: students engage in campaigns such as the collection of cooking oil and the wise use of water. For some of these campaigns there are partnerships with other institutions, such as the Dorina Nowill Foundation for the Blind where the school collects plastic lids that are sold by the foundation to generate income for their projects (Dual Stance School in the central region of São Paulo).
7. Clearances: the school leaves open spaces to be occupied, taking into account the concern with the environment, for example, a project to build a nursery (Living School in the south of the city).
8. Art and Environment: an environmental issue is addressed in art classes where recycled material is used to make toys, notebooks and stools (Ithaca College in the west of the city).
9. Adoption of plants: a school receives donations of seeds and seedlings that are 'adopted' by the students who take responsibility for caring for the plants and choose a location outside the school for planting them (Mobile School in the south of the city).
10. Goals: aimed at making the school more sustainable, the school created an agenda whose goals are reviewed annually by teachers and studied by students in classes in science and mathematics (IL Peretz School in the south of the city).

These can be seen in the Schools of Tomorrow Project that was created in 2009 and by August 2010 had worked with about one and half million people living in the *favelas* of Rio de Janeiro. There are 109,000 children and teenagers who attend 150 municipal schools that are affected by violence. The city of Rio de Janeiro has, for the first time, admitted that these students deserve special attention. They developed specific design workshops and introduced scientific experiments into the classrooms. A community educator was allocated to each neighbourhood to organise out-of-school activities and a director was appointed who needed to be a good manager, able to mobilise resources and make appropriate contacts.

There is a slogan that says, "Where there is a lack of education there is violence?" It can be argued that, "The school is the most important vector for sustainability, especially as an instrument of social aggregation, when it combines, for example, the elements of art and culture" (Berta 2010: 18). An example of this is the "Cientista do Amanhã" (Scientist of Tomorrow) Project that helps students, with teacher guidance, to make, for example, an aquarium. The "Escolas do Amanhã" have partnerships with volunteer mothers who are paid by the City Council. As for hiring trainees, university students who live close to the schools have preference in order to achieve a better identification for them with the local, everyday community in which they are working. By the time of the introduction of the "Escolas do Amanhã" Project a meticulous study had been undertaken to decide which organisations would take part in it. A number of aspects were analysed, such as the truancy rate, functional illiteracy, 'places at risk' areas, places with civil conflicts, places influenced by drugs trafficking or influenced by *milícias* (clandestine police). Of the 150 schools, 127 of them were located in very deprived communities and eight in communities with a very high level of violence, such as Rocinha, Vidigal and Morro da Dona Marta. There is also the "Bairro Educador" Project that was created to include the community in schools with a range of activities organised outside the schools.

A Small Scale Study

In a small scale questionnaire research study undertaken in April 2010 with students aged between 7 and 10 years attending eight schools located in areas influenced by trafficking, the students identified issues related to violence as the biggest problems in the places where they lived. In a school with 102 students located in an area controlled by a *milícia*, students referred to the lack of leisure options as their main worry (28.5%). Then they mentioned excessive garbage (17.8%). None of them referred to violence when answering the questionnaire. This was a municipal school located in a community occupied by a unit of the 'Polícia Pacificadora' (police officers appointed by the 'Secretaria de Segurança Pública' (Secretary for Public Security) to fight local violence. Their actions occur through the creation of links and partnerships with the community aimed at improving the quality of life of people who live there. The responses to the questionnaire from this school show that criminals are not the biggest worry for the students. When answering a question about the biggest problem in the slum, only 3.2% of a total of 87 students mentioned criminals, while 12.1% referred to the police. The term violence appeared in 17% of the questionnaires, followed by garbage, the biggest problem for 15% of the children. Drugs appeared in 5% of the answers and the presence of guns appeared in less than 1% of the answers.

The absence of violence in the answers of the students from this school located in the area with *milícia* does not mean, however, that the community lives in peace. Behind many stories, there is still the presence of drugs among relatives or the memories of familiar people murdered because of disagreements. But there is a difference between the schools located in areas controlled by trafficking: these places do not suffer from constant conflicts between criminals and the police.

Another example was a school where, despite being located in a place with a strong presence of drug dealers, violence appeared as the biggest problem in only 5% of the answers. The main fear, mentioned by 70% of the students, was the collapse of the hill on which they lived. The storm at the beginning of April 2010 left many people homeless in the same area in which their school was located.

In a second poll in the same period and at the same school, produced by the municipal government of Rio de Janeiro, with students of similar age, it was noted that they absorbed the everyday problems in the regions where they live. Of the 726 students from ten schools in areas of risk who responded, 407 (56%) cited problems related to violence. The 56% related to violence were divided into four categories, namely: violence generated by stray bullets or shots – 206 (29% of the total for the

Table 7.1 Students' responses to questions about violence and infrastructure

Problems	Number	%
Violence	**407**	**56**
Bullet shots	211	29
Bandits	131	18.1
Violence in general	46	6.3
Police	19	2.6
Infrastructure	**167**	**23**
Waste	72	9.9
Floods and landslides	61	8.4
Lack of water or light	34	4.7
Other issues	**152**	**21**
Total respondents	**726**	**100**

item), bandits – 136 (18.7%), general violence – 46 (6.3%) and police – 19 (2.6%). 167 students (23%) cited problems related to infrastructure, distributed as follows: garbage – 72 (9.9%); floods and landslides – 61 (8.4%); lack of water and light – 34 (4.7%). The distribution is shown in Table 7.1.

The municipal school system in Rio de Janeiro has 696,000 students in 1,062 schools. Of these, 109,000 students are studying at schools in the "Escola do Amanhã" (Schools of Tomorrow) Project.

The diversity of cultures present in the different zones of Brazil viewed from a foreign perspective is surreal. Because this diversity is closely related to the history of each group in each zone, it must be taken into account in schools. Brazilian culture is broad and rich: it has a diversity of social classes, of ethnic variety and a wide range of cultural expressions. While there are synthetic drugs available at the pharmacies, the popular medicine continues to use medicinal plants, the teas and the prays; while the media play American songs, the samba, the circle, the "catira" and the "acalanto" continue to be popular; while plastic and aluminum pots rule in the stores, stone and earthenware pots survive in many places in Brazil.

Environmental Arts

Since the 1970s environmental art, together with other new expressions, emerged with the utilisation of elements from the natural environment. In addition there are sculptures and performances that attract public attention to the problems of the world. Projects on these themes are necessary so that students can develop a consciousness of the necessity to embrace the cause of their common planet, its preservation and the rejection of any action that threatens environmental equilibrium. In this way, the arts class has an important role in communicating to students the seriousness of both the need to respect and preserve the environment and the need to recover what has already been destroyed.

A course titled "Museu de Arte Contemporânea" was organised by the University of São Paulo (MAC/USP) in 2008, directed at teachers and focused on the environment. It included debates, reflections and guided visits of students to museums. This way, teachers went to MAC/USP and participated in a variety of activities related to nature, starting from observation and games in the trees, moving on to an experience with the artist Octavio Ruth which created trees painted by them. Later they enjoyed the important scenery of sugarcane fields and coffee fields in the State of São Paulo painted by famous artists. The activity had three "knows": know how to do, know how to enjoy, and know how to reflect, clarifying the role of art in the construction of the knowledge of children.

Many artworks have extolled the diversity of nature in Brazil, depicting its flora, fauna and geography. In past centuries, artists have shown the purity of forests, rivers and blue skies, and even in a million years they would not have imagined that one day profligacy and pollution would take charge of that splendour. Among these artists, the Dutchman Franz Post painted several scenes. Albert Eckhout, a British artist, carefully painted the plants, animals, colours and customs of Brazil. At the time of Portuguese colonisation the French artist Debret documented the nobility, the slaves and the Indians in scenes full of trees, flowers and fruit.

While at that time it was the beauty of the natural environment that was shown, nowadays many works show the other side of nature: its destruction, environmental destruction, social destruction, political and cultural destruction. The German artist Baumganten painted on his canvas the extermination of the Ianomami Indians in the Amazon; the Chilean artist Alfredo Jaar painted with an emphasis on the miserable and inhuman life of the prospectors in the Brazilian forests, showing to the world the social, political and environmental problems informed by the visual arts (Barbosa 2008).

Together with those artists, we have the Polish artist Frans Krajcberg, a contemporary artist who lives in the State of Bahia. He recreates artistically elements of nature using different techniques and organic materials and minerals: twisted vines, plant fibres, trunks, colourful sands and pigments extracted from stones and the soil. The most important thing about this work is the collection of objects of nature – dead when collected, but with another meaning, changing their function and, through that, being transformed in art. The trunks and branches are from forest fires or they have been taken, already killed by parasites, from the mangroves.

An educational practice, different from paper and plastic paint activities, is the use of industrialised materials and this has been adopted by Lima (2007). Students aged between 7 and 10 years attending an elementary school in Sorocaba, in the State of São Paulo, collected materials such as leaves, seeds and branches that, because of the wind, rain or some other reason, were lying on the ground. They were transformed to a new aesthetics, becoming works of art derived from an ecological perspective. This was a form of environmental education that was itself political because it had a deeper meaning and sense.

Such experience, linked to life experiences, provides a theoretical and methodological foundation for educational practices where each person's experiences lead to citizen interventions. Art enlarges our perception and representation of the world and encourages us to deepen our theoretical and methodological knowledge to create an environmental education for planet citizens as Krajcberg wants. In his words (2005: 8),

> The fire is death, the abyss. The fire remains in me ever since. My message is tragic. I present the crime. I show that the addition of a technology is an abyss with no control. So I find the evidence, bring it and add all around my revolt that is the most dramatic and violent. This is how I burn and if I could put the ash everywhere, it would be closer to what I feel. I now work in my reminiscences of the war in my unconscious. Today actually works.

However, although creative and individualised attitudes are to be welcomed, progress can best be made through a fundamental and strong investment in the environmental education area, where interdisciplinary teams of trained professionals can act together with the schools working from their local realities in the communities from where their students come. However, in general, many problems still must be faced if these attitudes are to advance in the country. In addition to the deficiency in the production of educational materials for the development of an environment-oriented education, we face problems related to the cultural diversity of Brazil, the country's area (8,511,965 km^2) and the presence of a technology, such as the media, that is uncompromising. Indeed, each teacher is asked to respond to the challenges of education for sustainable development from the context in which it is introduced. Teachers have to take into account the particularities of each school, its neighbourhood, the social class of the neighbourhood, its institutions, social and cultural aspects and many other factors that demand the involvement and dedication of those involved who may not always have the necessary time available to address these issues.

Conclusion

The "Parâmetros Curriculares Nacionais" (PCNs, National Curriculum Parameters) were elaborated inside a group of policies aimed at supporting Brazilian educational reform, initiated by the "Ministério da Educação" (MEC, Ministry of Education) in the 1990s. They were published in order to guide schools in the elaboration of their educational policies, to aid them in the fulfillment of their teaching programmes and learning activities in classrooms and also to contribute to the elaboration and adaptation of educational schedules by the "Secretarias de Educação" (Secretaries of Education) from states and cities.

The MEC has used the "Constituição Federal" (Federal Constitution) of 1988 (Brazil 2006) and the "Lei de Diretrizes e Bases da Educação Nacional (LDB, Law of Directives and Bases of National Education), law number 9394", from 1996 (Brazil 1996) to make the PCNs. It seemed to be the intention of this policy of the MEC to provide a common national foundation. In short, the document highlights both the curriculum contents and competencies to be achieved by students and their teachers. The contents to be taught and learned in each area were designed to integrate information learned with attitude formation. The documents for each one of the areas start with a kind of explanation about the importance of the contents for citizenship formation and the practice of citizenship.

The PCNs were subjected to many criticisms when they were being made. These criticisms related to the allegation that the academic community had not been invited to take part in the drafting of the PCNs and to the conditions in which the discussion process had occurred. Critics pointed to the influence of Spanish curricular reform and international organisations in their development, reporting that decisions on the PCNs were centralised by international agencies and adversely affected the assessment systems, curriculum and even the system of redistribution of national resources.

The PCNs, as a curricular policy, were subject to some difficulties in their implementation. Academics, such as Ball and Bowe (1998), have pointed to a series of conditions that were essential in English schools for the application of a curricular reform proposal in their country. In Brazil, similarly, factors such as the availability of appropriate human and material resources in the schools, the interpretation teachers make of the PCNs, the personal and professional interests of the teachers and the groupings in current educational politics have influenced the use or lack of use of the Brazilian PCNs as guidance for the educational process.

Considering the PCNs as a textual and discursive product that results from intense processes of cultural and social negotiation that lead to new meanings, validations and denial of certain practices, there is a constant tension between the 'official' and the 'local or personal' proposals. However, this is not true for environmental education as teachers are aware, evidenced in several areas of activity, that environmental problems are not solved with sterile scientism, be it ecological, biological or technological. Resolution lies in culture, the social imagination, values, economic and political organisation, and local, regional, national and global perspectives.

Those who work in the environmental education area know that progress is slow and continuous. The overall aim is to change attitudes that are rooted in people's minds, with teachers and directors of educational institutes being no exceptions. It is known that the disregard of environmental questions occurs at all levels. As Carvalho (2003: 99) asserted, "through the national curriculum if it is possible to draw the attention of education professionals to the subject, leading them to self-evaluate and question their positions, efforts for their preparation will have been worth it". The United Nations Decade of Education for Sustainable Development 2005–2014 has drawn attention to the urgency of the changes necessary in the daily construction of sustainability in order to ensure that each individual plays his or her part in taking responsibility for building a better future.

References

Ball, S J, and R Bowe. 1998. El curriculum nacional y su "puesta en práctica": El papel de los departamentos de matérias o asignaturas. *Estudios Del Currículum* 1(2): 68–89.

Barbosa, A. 2008. *Tópicos e Utópicos*. Belo Horizonte: CI Arte.

Berta, R. 2010. Escolas para mudar o amanhã. *O Globo*. Rio de Janeiro, May 30, 2010: 18.

Brazil. 1996. Lei nº 9.394, de 20 de dezembro de 1996. Dispõe sobre as diretrizes e bases da educação nacional. Brasília: Diário Oficial da União, December 23, 1996.
Brazil. 1997. *Constituição da República Federativa do Brasil – 1988*. Brasilia: Brazil.
Brazil. 2002. *Parâmetros Curriculares Nacionais*, vol. 9. Brasília: Ministério da Educação.
Brazil. 2006. Constituição da República Federativa do Brasil, de 05 de outubro de 1988 (atualizado até a emenda Constitucional nº 53, de 16/12/2006). http://www.senado.gov.br/sf/legislacao/const/. Acessed on April 8th 2008.
Brugger, P. 1994. *Educação ou Adestramento Ambiental?* Florianópolis/Santa Catarina: Letras Contemporâneas.
Carvalho, V S A. 2003. Educação Ambiental nos PCNs: o meio ambiental como tema transversal. In *Educação Ambiental Consciente*, ed. C Machado. Rio de Janeiro: WAK.
Cordiolli, M. 1999. *Para Entender os PCNs: os Temas Transversais*. Curitiba: Módulo.
Dias, G. 1992. *Educação Ambiental: Princípios e Práticas*. São Paulo: Gaia.
Eckersley, R. 1992. *Environmentalism and political theory: Toward an ecocentric approach*. London: UCLA Press.
Krajcberg, F. 2005. *A Natureza de Krajcberb*. Rio de Janeiro: 6B Arte.
Leff, E. 2001. *Epistemologia Ambiental*. São Paulo: Cortez.
Lima, A. T. 2007. Frans Krajcberg: um cidadão planetário. *Revista de Estudos Universitários*, Sorocaba, 33(1), June, 35–148.
Macedo, E. 1999. Parâmetros Curriculares Nacionais: a falácia de seus temas transversais. In *Currículo: Políticas e Práticas*, ed. A Moreira. Campinas: Papirus.
Sanchez, C. 2003. Da institucionalização da Educação Ambiental. In *Educação Ambiental Consciente*, ed. C Machado. Rio de Janeiro: WAK.
Sancho, J. 1998. O Currículo e os Temas transversais: misturar água e azeite ou procurar uma nova "solução"? *Revista Pátio*, No.5, Year 2. Rio Grande do Sul: Artes Médicas, May/July.
Saviani, D. 2000. *Pedagogia Histórico-crítica: Primeiras Aproximações*. Campinas: Autores Associados.
UNESCO/UNEP (1977). *Intergovernmental Conference on Environmental Education, 1977*, Tbilisi, USSR. Final Report. CEI: Tbilisi.
Viola, E J. 1997. Contraponto e Legitimação (1970 a 1990). In *Ambientalismo no Brasil: Passado, Presente e Futuro*, ed. E Svisky and J P R Capobianco. São Paulo: Instituto Sociambiental, Secretaria do Estado do Meio ambiente do de São Paulo.

Chapter 8
School Culture in Brazil: Complexities in the Construction of the Sustainability Concept

Sergio Luiz de Souza Vieira

Introduction

When the Portuguese people arrived in Brazil in the sixteenth century economically supported by "Ordem de Cristo", a Portuguese organisation that helped the navigators, education was seen as a problem not as a solution. At first the religious missions tried to 'save the people', which meant the Indians, destroying the culture of those people living in the forests who knew how to teach for their survival in harmony with nature. The missions were to cause eugenic and ethnic changes and from then on the indigenous people were at the mercy of political interests exercising power in a centralised civil dictatorship. Currently, with no aims for national integration or making provision for gaining professional qualifications, education has become a tool for socialisation, unable to create the identification of students with the places where they live and, affecting in this way, the environment. The Brazilian school system continues in the twenty-first century to be aimless, while social projects have been introduced that are better welcomed by the population.

In this chapter we seek to analyse how this has come about, bringing out the key elements necessary for reflection about the school and its function to promote environmental education. Environmental education is absolutely necessary for the sustainability of future generations. The Food and Agriculture Organisation of the United Nations (FAO) has been alerting the world for many years to the possibility of the extinction of human life on our planet. They refer to the loss of the tropical forests, decreasing fresh water sources, a huge global population increase, environmental damage and, especially, the harm caused by greenhouse gases which may reach catastrophic levels. With all of this in mind, we shall discuss the ways in which ways education in Brazil may contribute to the application of sustainability

S.L. de Souza Vieira (✉)
Catholic University of São Paulo (PUC/SP), São Paulo, Brazil
e-mail: profsergiovieira@gmail.com

M.L. de Amorim Soares and L. Petarnella (eds.), *Schooling for Sustainable Development in South America*, Schooling for Sustainable Development 2,
DOI 10.1007/978-94-007-1754-1_8,

concepts and to guaranteeing for future generations the possibilities for a necessary equilibrium. The main focus of this discussion will be on basic education that includes elementary, secondary and high school levels. There will also be some brief references to special education.

Brazilian People and Their Educational Needs

'Official' Brazilian history started on April 22nd, 1500, when the Portuguese settlers arrived in Brazil with a fleet headed by Pedro Alvarez Cabral who came with a writer whose function was to register the 'discovery,' though this was eight years after the Tordesilhas Treaty. This Treaty between the Kingdoms of Portugal and Castela (today's Spain) and having England as a mediator was aimed at sharing the lands overseas so that there would be no conflicts between the two big Christian countries. The lands to the east would be Portuguese and the ones to the west would be Castela's (Spanish). Initially, it was called Ilha de Vera Cruz, then Terra de Santa Cruz and finally Brazil. This land was already known by this name before the arrival of the Europeans. In a way, this change from an island to a country with continental dimensions already pointed to the complexities, incongruities and contradictions which would be experienced in this country.

At first, the Portuguese did not attribute much importance to the land, because they were more interested in the wealth coming from the trade with their other colonies in the Orient. This way, the adventure called Brazil had, as a starting point, the initiative of two adventurers called Diogo Alvarez, also known by the nickname of 'Caramuru,' and João Ramalho, who came to Brazil on their own without the official authorisation of the Portuguese state. The first one arrived in Bahia and the second one in São Paulo and both started what we call today small Indian villages. These were places with Europeans and Indians having mercantile purposes. This happened through 'cunhadismo' which refers to marriage with an Indian girl from their village and with this arrangement it was possible to establish a closer relationship with all the men from that village, with mutual obligations such as the bartering of goods. This way, the men became 'cunhados' (brothers-in-law) and each one could ask the other what they wanted. For the Indians, utensils, arms and tools were required. For the Portuguese, the pau-brasil harvest was very valuable in Europe. To increase their profit, each of these adventurers married many Indian girls from different tribes, resulting in hundreds of children who were neither Indian nor Portuguese, but new people, who were reproducing indiscriminately and without limits. The distinguished anthropologist Darcy Ribeiro (1996: 106) called this process "construção da ninguendade" (the process of being nobody) since the ones who were born through cunhadismo lacked either a past or any relationship with a cultural identity drawn from ethnic roots and they had no feeling of being linked with the land. They were people who only had a future and they were denominated Brazilian people. And Brazil was really born from this adventure.

Considering that it was necessary to restrain both the pirates and the invading nations, Portugal needed to promote economic development in Brazil through a plan which focused on making people inhabit its territory. This plan was called 'capitanias hereditárias' that was a system where land was offered to persons close to the Portuguese Court, but it was unsuccessful because they did not want to abandon all the benefits and privileges they enjoyed near the power centre in Lisbon. However, it brought benefits for the economic cycle of sugar cane production that motivated African slavery in Brazil, since there were religious and cultural restrictions on Indian slavery. Consequently, from 1530 thousands of Africans arrived in Brazil and as had happened with the Indians the Portuguese had hundreds of children with the African women. So, the ones who were born were neither African nor Portuguese. They were 'nobodies' and, as time passed, they started miscegenation with the other 'nobodies' without any sin or crime but just being happy to increase the population. And so, this 'nobody process' produced, afterwards, the Brazilian people.

The history of education in Brazil started after 1549 through an agreement between King Dom Manual the 3rd and the *Companhia de Jesus* (Jesuit priests) and responsibility for it was given to the priest Padre Manoel da Nógrega. Today, this would be considered as outsourcing education as it was provided by the Portuguese State Agents. However, the changes proposed by Marques de Pombal that promoted the union between state and church led to the Jesuit priests being expelled from Brazil and in 1759 teaching became laic and the educational infrastructure was controlled by the Catholic Church, though this separation was not clear, since some religious orders and congregations had already incorporated this function. With the arrival of the Royal Family and its court in 1808, including ecclesiastical leaders, doctors, lawyers, bankers and college teachers, all fugitives from Napoleon Bonaparte, Brazil really started to experience economic, political and social development and in this way became the Portuguese headquarters.

At this time, Catholicism was the official religion and the Catholic Church had great power but it also had many obligations to the state. This meant that it had the functions of a registry, registering all the babies who were born, persons who were married and who died, where people were buried and to whom their belongings/properties should go. It also took care of people's health through Santas Casas de Misericórdia (Catholic hospitals where people with no financial resources could be treated) and congregations that also took care of sick people. Education was included among their responsibilities, and children from rich families could graduate into religious orders. There was also the care of the young and adolescents, preparing them for a real and worthy Christian life. So, Augustinian thought that dominated the centuries of Portuguese control condemned them to the eternal suffering of ideas: a person could choose between God or Satan, but the freedom lay not in choosing one or the other, but only God, because if Satan was chosen slavery and evil would result. So, the ideas became confused because a person was always trying to understand what was presented to him or her and having to identify if it came from one or the other.

Bearing in mind these circumstances, an acculturation process was established to resolve the conflicts of the three different ethnic groups that composed the Brazilian people: the Indians, the Portuguese and the Africans. This was seen as a negative factor, because miscegenation was considered to be a reason for sicknesses, people's weaknesses and moral disintegration. It was a time of constructing nation states and it was also a time when intellectuals were in endless debates. What was certain was that a successful nation state would be the one with the 'best people', which meant, the 'purest race'. Brazil was considered worthless to the world, a country without a future considering the origins of its people and so it was believed that Brazil was certain to be very unsuccessful.

Motivated by these concepts, the Aurea Law was signed on May 13th, 1888 putting an end to slavery in Brazil and the slaves were freed. However, this law also took from the state all the ssocial responsibilities regarding the population. The people were without protection, indemnities, education, health, work, housing, food, clothes and other necessities for their survival. With this law capitalism was implemented in Brazil. It was more of an economic plan than an action to save or benefit the Afro-descendents, as it was claimed to be. As a consequence, the construction of the first slums and housing in environmentally hazardous areas was commenced resulting in bad effects on the environment that have continued since that time. For each new slum there was violent repression in an attempt to reclaim the land. The Afro-descendents were abandoned to their own destiny, surviving as sub-employees: garbage collectors, grave diggers, street sweepers, washerwomen, maids, *mascates* (a kind of street vendor who sells cheap and popular objects), among others. The Government had a project to 'whiten' the cultural and biological aspects of the population, so Afro-descendents saw their jobs being taken by European immigrants, including Italians, Poles, Lithuanians and Germans.

On November 15th, 1889, Brazil became a republic and the persons responsible for this were the illuminists with their ideals of 'freedom, equality and fraternity'. However, surprisingly, at the last minute it was declared by the positivists, military and civilian followers of Augusto Comte's thoughts. This way, the state became laic and Augustinian thought which had led the monarchy was replaced by positivism, and the motto "Order and Progress" was inserted into the Brazilian flag. This expression of positivism meant that nothing could be skipped and everything needed to happen slowly following the natural order and resulting in progress. This was the starting point for racial discrimination against the Indians, blacks, the poor and women. These minorities lacked the minimal conditions necessary to decide the direction of the Brazilian nation. Its future would be based on scientific fundamentals. They lacked the competence to guarantee their citizenship. The Catholic Church, the official religion of the monarchy, also had State obligations for education, health and, in a way, justice, as it provided the birth registers, marriages, funerals and testaments but lost a considerable part of its power and influence.

As soon as the positivists took power, they faced a major challenge: the construction of the 'Nation State', where miscegenation was considered a degradable situation compared to the quality of the people and national ethic. Having this in mind, groups of intellectuals reached an agreement that the investments in education and health as governmental obligations would be important factors for the future of the Brazilian people. Its application was divided into the following three groups:

a. *Conservative* – They shared the opinion that the Brazilians were formed by the worst elements from each people, that is, the laziness of the Indians, the submissiveness of the Blacks and the stubbornness of the Portuguese. The only way to solve this problem was to import a national system of intellectual, moral and physical education from such developed countries as France, Germany, Switzerland and Britain.

b. *Nationalist* – They shared the opinion that the Brazilians were formed by the ones with the best qualities from each people who had arrived, such as the love of the land which came from the Indians, the strength of the Blacks and the willingness to open up the wilderness from the Portuguese. This way, they believed that Brazil did not owe anything to other countries and it should create a national system for intellectual, moral and physical education, once there were excellent pedagogues and physiologists.

c. *Vanguard* – They also had a nationalist direction, but they differed from the others because they believed there was no necessity to create a national system of education, but there was a need to improve what was already there: the people and their behaviour, their housing, their habits, their sexual relationships and all the undesirable cultural expressions, especially the ones of Afro origin. In order to improve, slums and *cortiços* (very small poor houses) were demolished, mandatory vaccines and sports were introduced. They were successful because, even today, we can see that some groups of blacks have become members of soccer teams, the samba originated in the Samba Schools, the *Congada* (a type of dance where the coronation of Congo's king is represented), small musical groups, *Capoeira* as a sports, *Candomblé* in *Umbanda* (religious beliefs), and so on.

The signs of this positivism are found even today when we hear speeches about dilettantism, public consciousness, social control and national jingoism as well as the speeches with a political approach, where real terms are replaced by more appropriate ones like: street people replaced by "people who are in a street situation", slum by "community", and underage by "child and teenager". This positivism had a great influence on national education but after this period the approach changed. Nationalist tendencies appeared in the Getúlio Vargas period (1930–1945) when a civilian dictatorship was established in Brazil. At this time, the nation state was consolidated and more concern was expressed for the workers, independent of their miscegenation, as well as for national industry and the creation of national companies in economically strategic sectors. Companies like Petrobrás, Instituto Brasileiro do Café (Brazilian Institute of Coffee), Fábrica Nacional de Motores (National Factory for Automobiles), the Siderúrgicas de Volta Redonda e Vale do Rio Doce (Volta Redonda and Vale do Rio Doce Metallurgy) started to emerge. Education had a utilitarian aim, to produce qualified manpower for these national production sectors, from the intellectual as well as moral and civic points of view, and also to care for the health of the workers.

In 1932 a group of educators was successful in summarising the aspirations of that time as an outcome of political and economic crises and the aspiration for educational transformation in Brazil. They communicated their views to the people and to the government in their *Manifesto dos Pioneiros* (Manifesto of the Pioneers). This publication emphasised that education should be regarded as a 'national problem' and presented a complex diagnosis as well as pointing out the direction which should be taken. It was an allegation as well as a cry of joy to formalise the educational policy based on a scientific plan. This document was the basis for statements about education in the 1934 Constitution, and it also resulted in an ante-project for the so-called 'National Education Plan'.

The plan was supposed to establish educational parameters that should be used all over the national territory, mainly regarding teaching through the Portuguese language, bearing in mind the construction of the national state. Because of this context, other institutions were created including the Education and Public Health Ministry and the National Council of Education. However, with another dictatorial *coup d'état* in 1937 a plan was implemented but "it denied all the theses defended by the educators from that philosophy" (Meneses 1998: 108). It was a centralised document which presented in detail the ideals of national education at all levels from primary education up to college level, including adult and vocational education, with enlarged curricular contents and evaluation criteria among others. The plans were introduced within a politically defined budget and education in its full sense was not a priority in the dictatorship.

With the 1946 Constitution, it was necessary to establish directions and bases for Brazilian education enforced by law. However, this did not happen either then or under Getúlio Vargas' successor, Juscelino Kubitscheck (1956–1961), even though it was a time of global planning for the government and state renewal.

From the Laws of Procedures and Bases of National Education

When João Goulart, who had a Marxist philosophy, became President of Brazil, Law 4024 was published in 1961. This was the first Law of Procedures and Bases for National Education (LDB). It was very distant from the *Manifesto dos Pioneiros*. New directions were given to education of which the main ones were: more autonomy for the state; a decrease in the centralising power of the Education and Culture Ministry; the introduction of the State Council for Education; the allocation of 12% of funds from the Federal Budget (Federal Government) and 20% from the municipal districts that were designated for education; the mandatory enrolment of students in the first 4 years of primary education; the establishment of criteria for the training of teachers for primary, elementary and high schools as well as colleges; the adoption of 180 days for the school year; and religious classes were not mandatory. All of these changes were decided and approved by the National Congress.

'Jango', the popular nickname for João Goulart, had tried to reform agrarian, fiscal, housing and educational sectors. With regard to education, Paulo Freire, a Marxist pedagogue, was asked to take responsibility for it. He wanted to see the end of all private and religious schools and make the state responsible only for the public schools and the political education of the citizens. This situation, as well as other reforms, was not accepted by most of the national political sectors. So, the press, the church, the OAB (Organ for Brazilian Lawyers), the entrepreneurs and several social classes started to call for a military intervention, demanding that the Marxists should step aside from their official positions. As a result, João Goulart was deposed by Federal Deputies and Senators and the presidency was considered vacant, as he ran away to Rio Grande do Sul and later to Uruguay. All of this contributed to the military taking power, establishing a governmental regime based on bipartisanship and the indirect vote that led to a succession of five military general presidents from 1964 to 1985. During this period, some actions were taken to create permanent aims: to bring Marxist threats to an end, to promote economic and social development as factors for security and national integration and improvements in education as steps to achieve these goals successfully.

During this period, when General Emilio Garrastazu Médici was the president, the second LDB was issued in 1971, through Law 5692, which tried to consolidate the national state while keeping some of the changes already approved in 1961. It also brought some important innovations, including the implementation of a common content for the curricula of primary and elementary schools and curricular diversification according to particular local circumstances. In addition it sought to maintain moral and civic education as well as physical education as created in the Getúlio Vargas government. Arts education and health programmes were made obligatory subjects in the curriculum for the 1st degree level – for students from 7 to 14 years old. Distance learning was offered as a possibility for the 'supplementary teaching' (a kind of course, where you can take the elementary and high school levels in less time than usual). Special graduation was introduced for teachers of the 1st degree – from the 1st to 4th levels, but also having training for the 2nd degree. There was special graduation for the teachers of 1st and 2nd degrees with higher education graduation for experts at the higher education level or post-graduate levels. Public money was made available not only for public institutions. It was determined that the municipalities had to invest 20% of their budgets in education.

This Law was considered by the progressives to be responsible for worsening the education system because "it eliminated any possibility to establish policies, plans and education as effective tools for a desirable development to the Brazilian education" (Meneses 1998: 111). However, public education was considered remarkable because of the high quality of what was offered to the population through all schools, which meant teaching democratisation. An interesting aspect was that those students who were unable to benefit from free and public schooling had to try to find private schools for additional studies. This situation occurred with students who could not succeed in public schools. This meant that if a student had failed twice in the same level he or she would have to leave the public school. As a consequence there was an effective participation from the family, parents, siblings and others in a student's performance.

Currently, this kind of participation does not occur. In general, families have delegated to the state the mission to educate their children and this has worsened the social situation. There was a great paradox. Subjects such as music, philosophy, social organisation and Brazilian politics, moral and civic education, social studies, industrial arts, arts education, study of Brazilian problems and public health besides French which was also taught with English, have contributed to the development of more extensive competences but are not included now in the curricula of public schools.

In that historical period, education was given high priority because it was considered a factor of great importance for social integration and national security. Consequently, the public schools received special care from the military government. It is important to mention that at that time, besides nationalism, liberalism led Brazil to high levels of economic development, which was referred to as the "Brazilian miracle", precisely at the time when the world was in a period of crisis.

With the re-democratisation of the country, resulting from the political opening up and the exiles' return, possibly due to the political amnesty, and mainly following the election of civil presidents, it was mandatory to edit again the Constitution of the Federal Republic of Brazil. This took place in 1988. As a consequence, education had to be restructured and this became effective through Law 9.394 from 1996, under the governorship of President Fernando Henrique Cardoso (1995–2003). It also established the present Guideline Law and Bases for National Education, also called the Darcy Ribeiro Law, named after an important anthropologist who had been an active participant in the present formation of Brazil.

The current LDB has brought the following innovations:

- democratic management of the public education system and progressive pedagogic and administrative autonomy for the scholastic units;
- elementary education was made mandatory and free;
- for basic education, the school year comprised a minimum of 800 hours distributed over 200 days;
- designation of a common core plus a diversified element, to take account of local particularities, for the curriculum of elementary and high schools;
- for teachers wishing to work in basic education, teacher training was provided in higher education courses;
- trainee teachers for infant schools and the first 4 years of elementary schools graduated in a "Normal course";
- for high schools, specialist teachers graduated in pedagogy or in post-graduate courses;
- the Federal Government should spend at least 18%, and the states and the municipality 25%, of their budgets to maintain and develop public education;
- public money could be used to finance the community, religious and philanthropic schools;
- the establishment of the "Education Decade" as a "National Plan for Education."

This LDB also established the PCNs (National Curricular Parameters) in each subject to be taught and these were absolutely mandatory at all levels of elementary and high school education and in all public, private, religious and community schools. These PCNs are a set of proposals about the organisation and development of the school curriculum and they are an important reference for all schools, heads and teachers. They are neither an imposition nor are they homogeneous; they are flexible and open. They are parameters to obtain a satisfactory level of national education without losing the focus on the different socio-cultural realities of individual schools as well as the regions in which they are located. This situation is aligned with serious Brazilian social issues together with the emergence of environmental problems that have impacted on all nations. Discussions were conducted in schools through educational cross-curricular themes.

Challenges to Sustainability

The last decade of the twentieth century witnessed serious global concern for changing climate conditions and their effects on the environment and especially the impact of greenhouse gases on the global community. The United Nations Food and Agriculture Organisation (FAO) has been alerting governments to the imminent problems related to worldwide shortages of food supplies resulting from such circumstances as: deforestation, pollution of seas and rivers, the shortage of drinking water, soil erosion, global warming because of the greenhouse gases, the non-total absorption of carbon monoxide, the lack of alternative sources of energy, rising sea levels due to the melting of the glaciers and polar regions, as well as other harmful factors that threaten the survival of human life on our planet.

This concern has demanded deep international reflections about the necessary parameters to build a sustainable society and a sustainable planet. A number of international, inter-governmental conferences brought together scientists, world leaders and representatives of non-governmental organisations to try to reach solutions to the environmental and sustainability problems. Such conferences were held in Rio de Janeiro (Eco–92) in 1992 and in Kyoto in 1997 where attempts were made to agree a protocol to reduce carbon emissions. More recently, similar international inter-governmental conferences have been held in Copenhagen in 2009 and Cancun in 2010. The impact of these and other international conferences can be detected in individual countries and in schools. In the Brazilian education system, scholastic debate about sustainability was introduced in the cross-curricular themes foreseen in the National Curricular Parameters. This provision was appropriate because global problems demanded answers from worldwide citizens and, to make this a reality education was seen as having a very important role.

These cross-curricular themes tried to stimulate reflection as well as to guide discussions about common issues across subjects in the school core curriculum. The themes were chosen according to the social diagnosis of the time. They took into consideration: urgency regarding social affairs, the problems encountered all over the national territory, the possibilities for teaching and learning strategies according to each student level, facilitating an understanding of contemporary reality, and the acquisition of citizenship. As a first step the following themes were introduced:

Ethics

This is understood here to refer to the "collective good". It has as its main objective encouraging students to think and behave according to the question "How to act and behave when facing others?" This subject is very complex because it does not have just one rule, but it "implies that we should place ourselves beyond the static concept and go into the macro-concept space…A reality which is not defined any more by the thought of the object, but by the system-organisation concept" (Pena-Vega and Almeida 1999: 92).

Environment

This concept does not apply only to the ecological equilibrium between fauna and flora, it also includes the habitat which means care for the places in which we live. This terminology has been expanded as a result of the effects caused by the greenhouse gases and the local and national environmental problems that have also become global problems.

Cultural Diversity

This is a problem which challenges nation states, especially nations in the West with an illuminist philosophy; once they were based on equality legally established, but they have enacted more and more legislation that guarantees the right of people to be different and equal.

Health

The concept of health refers not simply to notions of sickness, but also, according to the World Health Organisation, to a search for improved general welfare, mental, physical and spiritual. To achieve this and to reduce ill-health, a number of factors must be taken into account including: sanitation, access to education, better income distribution, consumption of healthy foods, psychic equilibrium especially in urban environments, and habits and hygiene.

Sexual Orientation

This must be one of the most serious contemporary problems. There has been an increase in family breakdown in Brazil resulting in a loss of references about sexual education. Added to this, there is the continuous presence of infantile eroticism stimulated by television and by commercially driven consumption habits. As a result, we have an alarming rate of increase in the numbers for early pregnancy and this has become an important health problem leading to some schools opening their own day care centres intended to educate young mothers.

These themes were considered really important and they have raised the possibility of others to come. They served to combine three subjects – work, consumption and citizenship – in one. However, this has not produced a positive result since they were not officially deployed. There was the necessity to discuss the freedom which "does not identify itself with the way man is, but it is a fundamental part, together with other fundamental parts, too, as life, thought and work" (Mondin 2005: 121). And so, the state was concerned only with the conventional subjects – Portuguese language, mathematics, history, geography, physics and biology – in the way that the school could be used as a tool to improve society. The initiative was very welcome among heads, teachers, parents and educational thinkers. However, this was not enough since the political, economic and different social conditions led to an undesirable situation.

To help to clarify this issue, a preliminary questionnaire study was undertaken in March 2010 with 50 teachers from two schools, a primary school and a secondary school located in the East Area of São Paulo City. One school was near rivers and streams where slums had been built. Many of the people survived by recycling material and what was not usable they threw into the streams. The other school, on the other hand, was located in a more affluent area.

Our findings showed that, in general, the teachers deal with the educational cross-curricular themes in their subjects but they also engage in extra-curricular activities where they can work on sustainability themes without causing any harm to the development of the specific contents of their regular subjects. It was noticed that, in a way, the students could learn some general principles from these themes. On the other hand, it was observed that they had great difficulties in applying their learning in their own homes, which means they obtained relevant knowledge but could not use it in their own environment. This led us to a search for the reasons why they had these difficulties.

In our analysis focused on this new problem we could detect that these students do not care about the place in which they live, so they do not notice the environment where they are located. In this way, they do not have any links with the community which, for them, is simply temporary and meaningless. For the ones who live in the poorest regions, in general, schooling is not seen as offering an opportunity to allow a better life but as an institution which supplies immediate social benefits, such as: *Fome Zero* (Zero Hunger), *Renda Mínima* (Minimum Income), *Bolsa Família* (a help the federal government provides for the poorer ones), *Leve Leite* (the state provides milk for young children who belong to families with very low incomes), school meals and sometimes clothes which are given thanks to the efforts of teachers and employees. In a general way, these programmes occur in the three sectors of government: federal, state and municipal. They try to overcome economic and social inequality with emergency measures to promote individual emancipation. Some of these programmes used to provide food but now they supply the equivalent in money. This changes the relationship of the student with the school which becomes only a way to obtain financial subsistence for the family and not a place for the student to be trained for work and the improvement of society. Some critics say that due to the structural poverty which exists in Brazil some of these programmes have stimulated a baby boom especially among the poorest citizens. For each child the family income rises and this increases the poverty. The programme Minimum Income intends to increase to R$400.00 the value given to each Brazilian for every child. So, for millions of families the number of children they have might determine a better income than if they worked. For some it will be more profitable to have children than to have a job. The improvement of life without a responsible basis will result in higher consumption and the search for comfort and consequently the search for more products, raw materials and energy sources. All of this will accelerate the end of the sources for life preservation. Clearly, it is necessary to give a second thought to sustainability for social programmes and not only those with political purposes. The common subject of conversation among the poor is the possibility of moving to another place "where life becomes better". This way, according to our understanding of the world, their temporary place will always be the place receiving minimal government attention.

The ones who live in more affluent areas seemed to be more worried about the environment and the improvement of their social condition, but their wishes for social mobility, acquisition of new technologies and the search for comfort did not allow them to notice the environmental impacts caused by their life style. It can be understood as a life of comfort and social climbing, "the way men disrespect their own limits and value their hypertrophy" (Morais 1992: 86). It is a depreciation of their values created by the low value given to things. According to Morais, the excess of comfort is closeness to death, because it diminishes human longevity. It also causes a shortage in the resources and energy sources necessary to maintain all this unsustainable luxury.

It was noticed, too, that teachers generally speak about the existence of 'local power' but not as an institution where the central state has a minor participation. It affects people's lives through the dealing of drugs and arms, dominating the recycling of material and waste, the control of illegal and alternative transport, corrupt trade in items obtained from social benefit programmes, political control of housing even when it is very poor, and other actions for informal neighbourhood leadership. The applicability of the cross-curricular themes to these communities is considered null because of these facts.

We can observe that schooling as the only factor promoting and developing sustainability is insufficient. It is necessary to have strict action from the state to guarantee that the content can be taught effectively in schools. This can be accomplished not only with education but also through social justice, better income distribution, job creation, sanitation, security, good quality transportation, better health conditions and an effective housing policy.

Findings from the two locations in our study were confirmed by the teachers from the Pedagogy Courses taught in two universities located in São Paulo City. Although the two locations should not be regarded as typical Brazilian examples, they demonstrate the interfaces between the educational cross-curricular themes and the local culture, which vary according to local circumstances.

The state needs to guarantee that education is a right for everybody. When this does not happen, a person suffers from the lack of access to the process of education, which also results in difficulties in conserving the environment. Attempting to achieve sustainability becomes meaningless once these basic requirements are not observed.

It needs to be emphasised that the sustainability concept is organic, dependent on the harmony between those social, economic, cultural and environmental diversities that involve society in a general way. As a consequence, it demands a very high level of quality in popular education, especially for students of school age, bearing in mind the need for them to acquire appropriate habits. If all these are not fulfilled the aims of basic projects will not be reached. These projects are established having in mind the acquisition of what is ecologically correct, economically possible, socially fair and culturally accepted or assimilated.

The insertion of educational cross-curricular themes in the National Curricular Parameters established by the most recent Guidelines Law and Education Bases had as an objective the reaching of this ideal state. It was necessary to coordinate knowledge in a systematic way. This knowledge needed to be classified according to the arguments and national theories for different existing realities but in an integrated way. Ethics, environment, cultural plurality, health and sex education, as cross-curricular themes, are at the forefront of this integration. This integration could be achieved in inter- and trans-disciplinary ways, allowing for a complex knowledge acquisition which was supposed to be present in all phases of basic teaching.

So, the situation for education in Brazil at the beginning of the twenty-first century shows how serious the problem is and it also shows us how far we are from the ideal of sustainability. We would suggest that the current situation is unacceptable.

The Present Scenario of Education in Brazil

Here we shall analyse figures supplied by official agencies, having as our aim to look for some projections of questions related to sustainability obtained by high quality education. At first (Table 8.1), we have some data provided by the Development Institute for Basic Education (IDEB).

Even though the growth rate for the Brazilian population, as presented by the Brazilian Institute for Geography and Statistics (IBGE) was 1.00% in 2007, 1.2% in 2008 and 1.19% in 2009, pointing to a current total of 192,764,264 people, we notice in Table 8.1 a decrease in the provision of some types of schooling, such as special education, which was reduced by 21.01% and elementary schooling with a negative rate of 1.18% in 2009. With regard to enrolment rates, these varied according to the governmental budget and the population growth rate (See Table 8.2).

It can be seen in Table 8.2 that although the budget approved for the Ministry of Education has been greater the expected results have not been achieved. The majority of students enroled in elementary schools do not go on to enrol in high

Table 8.1 Changes in public school enrolment in the different areas of basic education in Brazil (Source: INEP 2010)

	Enrolment in basic education				
Phase/method	2007	2008	2009	Difference 2008–2009	Variation 2008–2009
Kindergarten/nursery school	6,417,562	6,179,261	6,672,631	+493,370	+7.98
Elementary school	31,733,198	32,086,700	31,705,526	−381,174	−1.18
High school	8,264,816	8,366,100	8,337,160	−28,940	−0.34
Vocational course	682,431	795,459	861,114	+65,655	+8.25
Youth and adults teaching	4,940,165	4,945,424	4,661,332	−284,092	−5.74
Special education	304,882	319,924	252,687	−67,337	−21.01

Table 8.2 Changes in enrolment rates and enrolment budgets (Source: INEP 2009)

Enrolment budget	Elementary school	High school	Difference	Variation %	Budget billion R$	Population growth
2007	31,733,198	8,264,816	–23,468,382	–73.95	27.02	+1.00
2008	32,086,700	8,366,100	–23,720,600	–73.92	31.20	+1.22
2009	31,705,528	8,337,160	–23,368,368	–73.70	41.50	+1.19

Table 8.3 Changes in the number of schools providing basic education (Source: INEP 2010)

	Schools for basic education				
Phase/modality	2007	2008	2009	Difference 2008–2009	Variation 2008–2009
Kindergarten/nursery school	110,782	113,550	114,158	+608	+0.53
Elementary school	154,321	154,414	152,251	–2,163	–1.40
High school	24,266	25,389	25,923	+534	+2.10
Vocational course	3,230	3,374	3,535	+161	+4.77
Youth and adult education	6,978	6,702	5,590	–1,112	–16.59
Special education	55,217	42,018	40,853	–1,165	–2.77

schools. The most important reason for this is that the majority of social projects directed to the underclass are linked to enrolment in elementary schools but not to enrolment in high schools. This suggests the following: the enrolments in the first years occur more for social benefits than for educational benefits; the high drop-out rate in high schools shows that the students do not see the school as an option to improve their *status quo*. It can be seen in Table 8.2 that there will be very serious social problems in the future because the students are not qualified for work and they will continue to be dependent on new social programmes.

As can be seen in Table 8.2, 2009 was the year when most resources were directed towards the Ministry of Education – 41.5 billion Reais (Brazilian currency) – whereas the development of basic education in Brazil seems to have stagnated. This increase does not mean that education, a crucial sector in the country, has received more attention from the government. It only represents the result of a larger receipt of federal income. In 2003, the beginning of the present government, the percentage directed to education was 2.88% of the budget. In 2004, it was reduced to 2.6%, in 2005 it was scaled down even further to 2.44% and in 2007 it returned to the present figure of 2.87% which is less than that of the previous government.

Turning to the number of schools in Brazil, as Table 8.3 shows, this situation is also critical.

In Table 8.3, it can be seen that there was a considerable reduction in the number of elementary schools. The growth in other sectors is also very poor. It is necessary to highlight the reduction of youth and adult education as well as special education, with a remarkable decrease in the last 3 years.

Table 8.4 The IDEB: changes in the grades obtained by basic education students in "Provinha Brasil" (a kind of assessment test) (Source: INEP 2009)

Evaluation – IDEB	2005	2007	2009
Public schools	3.6	4.0	4.0
Private schools	5.9	6.0	6.3
Elementary school: average	3.8	4.2	4.2
High school: average	3.4	3.5	3.5

Table 8.5 Student enrolments in private schools (Source: FNEP/FGV 2009)

Enrolments in private schools	Basic teaching	Elementary school	High school	Dropout	%
1999	6,648,568	3,277,347	1,224,363	2,052,984	–62.6
2005	7,019,189	3,376,769	1,097,598	2,279,171	–67.4

The numbers without qualifications for work in Brazil must be emphasised and, as a consequence of this, structural unemployment makes the government promote policies for entrepreneurship without taking into consideration the sustainability guidelines that promote incentives for handicraft production. Entrepreneurship stimulates the extraction of raw materials from the tropical rainforests to supply and meet the demands from tourism, sports, 'folkloric' and religious bodies. This governmental action exceeds the necessities for environmental preservation. These unsustainable practices become stronger considering that the contradictions are part of the Federal Constitution of Brazil. The traditional people have their ways of 'knowhow and do' stimulated, allowing them to cut down trees from the tropical rainforests aimed at supplying timber for handicraft work and this conflicts with the rules for environmental preservation.

Table 8.4 shows the average of the results obtained by students in the evaluation systems which are related directly to the kind of teaching offered.

Even though the results obtained by students of private schools are higher, the percentage of the students enroled in these schools is very small compared to the percentage of public school students. According to a study carried out by the Getúlio Vargas Foundation (FGV) for the National Federation for Private Schools (FNEP), there are also private schools with different development levels. The findings are shown in Table 8.5. Here we can see that in the private schools there is a big difference between the enrolments in elementary schools and high schools. The dropout problem in private schools is shared with schools in the public school sector.

Table 8.6 presents figures derived from the Evaluation System for Basic Education (SAEB) pointing out the quality level percentages/achievement level percentages found in the assessments for students in public and private schools based on two subjects considered to be basic: Portuguese Language and Mathematics.

The difficulties that exist in private schools are remarkable. These data should also be compared to the ones presented by the Evaluation Programme for Educational Systems (PISA) which indicates that Brazil is ranked 54th among the 57 countries

Table 8.6 Assessment levels in private and secondary schools (Source: SAEB/INEP 2010)

	Portuguese language		Mathematics	
Qualification	Private	Public	Private	Public
Adequate	27.6	37.0	27.7	2.1
Intermediate	59.9	56.6	53.7	34.7
Critical	10.8	34.0	17.2	53.3
Very critical	1.7	11.6	1.5	9.9

Table 8.7 Changing illiteracy rates in percentages 2005–2008 (Source: INAF 2009)

Functional illiteracy	2005%	2007%	2008%
Illiterate	11	9	7
Functional illiterate	26	25	21
Functional literate	63	66	72

evaluated in reading and 49th among the 56 countries evaluated in mathematics (INEP 2009) It would appear that the state is having difficulty in promoting the importance of education for the development of the individual. Once again, it is unlikely that the effective application of sustainability aims will be achieved. It is a structural problem and recent governments have been unable to solve it.

An important consideration is provided by the Office of Basic Education Development (IDEB), which assesses through 'Came from Brazil' (a test) the level of education achieved by students. This relates both to a certain stagnation of pupils in primary and secondary schools, because it was detected that the Brazilian students in 2007 had an average of 3.5, which is well below the average for developed countries of around 6.0 (INEP 2009).

In Brazil, private high schools offer better quality education and so students in these schools are better prepared for entry to those universities considered to be the best in Brazil. These universities enrol only 8% of students in the country and this is far from the goal established by the National Education Plan of the Brazilian Government. The public high schools do not provide a good quality of education and students have access to alternative private universities lacking structure and without quality education. Private universities cater for 92% of Brazilian students.

Another important factor in the analysis is that although there are social programmes for the inclusion of public school students in public universities, sorted by qualifying examinations, these programmes do not include these students. Universities face the problem of having to reduce their level of quality education to include these students. We have, as an example, the grades achieved in the National Examination for Performance Evaluation (ENAD) that is applied to college students in which no one has been awarded the maximum grade.

We still need to analyse the functional illiteracy percentage concerning the Brazilian population aged between 15 and 64 years shown in the figures produced by the Instituto National de Analfabetismo Funcional (National Institute of Functional Illiteracy: INAF) in Table 8.7.

Table 8.8 Illiteracy rates by level of education (Source: INAF 2009)

Functional illiteracy	1st/4th grade %	5th/8th grade %	High school %	Higher education %
Functional illiterate	54	24	6	1
Functional literate	46	76	94	99

In addition to the functional illiterates who constitute 21% of the Brazilian population, many who know barely how to read and write are unable to understand a simple text. This is a very serious problem. However, the situation is still more alarming, if we consider the population who are at school.

Considering the figures in Table 8.8, we have to try to find out the consequences of the 54% of students with functional illiteracy in the 1st to 4th grades of elementary schools. An obvious effect is to be seen in students' inability to engage in reading comprehension, getting worse at each grade and continuing to the end of high school. The rate of functional illiteracy decreases but the necessary knowledge is not acquired. In this case new epistemologies of education should be offered. It must be that the knowledge the person brings with him or her could be used and changed to something valuable for the community itself. This means changing the pedagogic basis. Some of the problems encountered by teachers in schools arise because they see the students as a group needing to be improved rather than a group with its own particular difficulties and with its own collective potential. "This is what is in question in post-modern changes...in order not to stifle the social aspect" (Maffesolli 2004: 16).

Having been presented with so many official aspects of education in Brazil, these need to be discussed according to the sustainability parameters, in order to analyse the extent to which schools are responsible for building up and implementing sustainable habits in their students, hoping that as global citizens they can preserve the planet as well as its energy, life and raw material sources for future generations.

Conclusion

As we have seen, Brazil started from the transgression of two Portuguese who, without the permission from the Portuguese state, initiated an adventure that resulted in miscegenation with the Indians thus generating a new people. As time passed and with the increasing necessity for economic development, Africans were brought to Brazil, who, through the miscegenation with other ethnic groups, changed the human scenario with the mixing of people of different colour. This process was known as the process of 'nobody' since none of these peoples could keep its own culture. The only way to change this situation was with the help of education. On the other hand, Portugal gave the Catholic Church the responsibility for education but only the children from the elite were entitled to receive it.

In a second period, this time as a positivist Republic, the Brazilian government denied citizenship to African-Brazilians. They were abandoned by the State, and had no access to education, health provision and work. The government tried to wipe out these people and their culture. This process has caused deep wounds that continue today and the associated social problems in Brazil are far from resolved.

Important actions were initiated by intellectuals and pedagogues who tried through a "Pioneers' Manifesto" to resolve this serious situation. However, the dictatorial government of Getúlio Vargas directed the educational structure of the country to his political purpose and this meant holding on to power and so the pioneers' recommendations were abandoned and forgotten. Something positive that happened in this period was the creation of the Ministry for Education and Public Health as well as the National Council for Education.

When the military regime was in power from 1964 to 1985 there was some progress in education, in its different levels, because it was considered as of fundamental importance to the economic development of the country and especially for national security. From this viewpoint, it was utilitarian education following strategic criteria and not aimed at the individual as a priority.

After this period, there was what we call 'country re-democratisation' which was not exactly a re-democratisation in terms of education because all its problems continued to be unsolved and there was stagnation in critical levels and in some cases there was a regression. While minor improvements could be felt, however, they could not be considered meaningful.

In 2007, there were alarming figures for enrolment in public elementary schools (31,705,528 students), in public high schools (8,337,160), in private elementary schools (3,376,769) and in private high schools (1,097,598) but in higher education the figure was only 7,136,354 and these results represent social stratification, considering that high dropout rates are still a reality. Social programmes included in public policies stipulate that the benefits received depend on enrolment in elementary schools. This way the students started to acquire another kind of link related to school. The extreme poverty factor and the factor of being disqualified for work have produced an erroneous view of the reality of schooling. The school now is seen as a place where students satisfy their hunger (hunger for food), not their hunger for social improvement through education. Going to school does not mean acquiring knowledge to change their lives, but attending school to keep their bodies alive. This is a paradox because in a wider sense the school should be the best way to save human life on the planet. As long as this management culture considers education simply as a legal obligation from the state and the family and not as an ethical, moral, social and environmental obligation changes cannot be foreseen.

This situation leads towards a lawless society. In general, families in Brazil do not provide the ethical fundamentals: schools do not educate, churches do not indoctrinate, health care does not cure, the banks do not work, the political parties do not politicise, the police do not police and the government does not govern. These are the usual consequences of the negative results of education. Organised and petty crimes definitely contribute as the student does not establish any relationship with the place where he or she lives. The rates of illiteracy and functional illiteracy are still high. There are those who can only read an address, street names, prescriptions, the bus or subway fare and other similar information. Considering these factors and the social ones which demonstrate the extremely low levels reached by the students such as reading in their own language and also mathematics, compared to the evaluation rates from better developed countries, forces us to recognise how critical our situation is in a global context.

However, we need to recognise the efforts which have being made by the authorities, heads, teachers, parents and organised civil society in trying to reach better solutions. But this is not enough! It is necessary to adopt drastic changes to the way problems are analysed and resolved since the educational structure in Brazil has been shown to be defective for many decades. Bearing in mind the measures necessary to implement strategies for sustainability and the figures presented in the Tables earlier, it is clear that the present model of educational administration in Brazil does not provide the conditions necessary for the population to adopt the sustainable habits essential for the next generations.

Finally, depending on the poor results from education in Brazil, which have always been considered as a matter of less importance, and, according to the ideologies adopted by successive governments, not as a matter for the state, there has been an uncontrolled exploitation and extinction of the bases for human life on our planet. It is considered unscientific to demonstrate emotions in an academic publication but as an active member of the global society, one would have liked so much not to have reached this conclusion and also not to have written this paragraph.

References

FNEP/FGV. 2009. Números do Ensino Privado. http://www.fenep.org.br/pesquisafgv/relatorioBrasil.pdf Acessed 9 Feb 2010.

INAF. 2009. Indicador de Alfabetismo Funcional http://www.ipm.org.br/ipmb_pagina.php?mpg=4.01.00.00.00&ver=por Acessed 10 Feb 2010.

INEP. 2009. Índice de Desenvolvimento da Educação Básica. http://portalideb.inep.gov.br/index.php?option=com_content&view=article&id=47&Itemid=57. Acessed 12 Oct 2009.

INEP. 2010. Sistema Nacional de Educação Básica. http://provabrasil.inep.gov.br/index.php?option=com_content&task=view&id=81&Itemid=98. Acessed 6 Feb 2010.

Maffesolli, M. A. 2004. *A Parte do Diabo: Resumo da Subversão Pós-Moderna*. São Paulo: Record.

Meneses, J. G. C. 1998. *Estrutura e Funcionamento da Educação Básica*. São Paulo: Pioneira.

Mondin, B. O. 2005. *Homem, quem Ele É: Elementos de Antropologia Filosófica*. São Paulo: Paulus.

Morais, R. 1992. *Estudos de Filosofia da Cultura*. São Paulo: Loyola.

Pena-Vega, A, and E P O Almeida. 1999. *Pensar Complexo e a Crise da Modernidade*. Rio de Janeiro: Garamond.

Ribeiro, D. 1996. *O Povo Brasileiro*. São Paulo: Companhia das Letras.

Chapter 9
Schooling for Sustainable Development: Experience with Students of Early Childhood Education in the State School Paulo Tapajós, São Paulo/Brazil

Maria Lucia de Amorim Soares, Leandro Petarnella, and Eduardo de Campos Garcia

With regard to education for sustainable development, there are two fundamental questions: What is sustainability? What is the role of schools in preparing young people for sustainable development? To answer these questions, it is necessary to consider some important principles. First, people as a subject are historically constructed. The historical context affects how people relate to their surroundings just as the society in which people live is also influential. This means that our daily practices consist essentially of a knowledge formation process. Similarly, these same practices will be responsible for forming the knowledge of future generations. Secondly, this knowledge is transmitted through the teaching and learning processes. As a result, we can only avail ourselves of the knowledge we have inherited from past generations through this process, education. Education is responsible for the construction of knowledge and the development of future learning. This implies that education is an activity with essentially human and humanising values. These considerations indicate that education has a decisive role in human life. It is responsible not only for the development of practical actions but also for *Praxis* development. Practice, as a technical activity, aims to develop the skills and competencies of people in preparation for entering the job market. *Praxis* refers to the exercise of thinking about one's own practice (practice thought out) relating to knowledge and reasons for action.

M.L. de Amorim Soares (✉)
University of Sorocaba (UNISO/SP), Sorocaba, São Paulo, Brazil
e-mail: maria.soares@prof.uniso.br

L. Petarnella • E. de Campos Garcia
University Nove de Julho (UNINOVE/SP), São Paulo, Brazil

M.L. de Amorim Soares and L. Petarnella (eds.), *Schooling for Sustainable Development in South America,* Schooling for Sustainable Development 2,
DOI 10.1007/978-94-007-1754-1_9, © Springer Science+Business Media B.V. 2011

Bearing in mind the principles set out above, as a starting point, we turn to an understanding of *what is sustainability?* In addition, *what is the role of the school in preparing young people for sustainable development?* The period of the industrial revolution was distinguished by the formation of large urban centres, a rural exodus, consolidation of the dominant capitalist system as a model and the advancement of technology to the detriment of human skills. The industrial revolution initiated a period when social organisations came to understand how economic systems generated wealth. In turn, it became important to emphasise that "all industrial systems tend to growth, and the whole mass production intended for consumption has its logic, which is maximum consumption" (Morin 1969: 37), which is driven by consumption now, because it relies on immediacy. It has been acknowledged that, through the industrial revolution, *human* relationships were transformed into *marketing* relationships and people – until then the owners of their own production processes – were transformed into a labour force, becoming part of the finished products. It is from this revolution that technology overlaps with nature. This implies that the "industrial revolution brought with it the foundation of a new spatial segregation, a new theory of value and a new reality for the law of value" (Santos 1994: 7). This is how the industrial revolution led to a transformation in both the way in which people saw themselves and in their relationship with their surroundings.

The logic of value, the guiding axis in the people–nature relationship, became the internal structure of nature and no more was nature defined as the human environment. The construction of knowledge, from this perspective, will unfold in the construction technique of transforming nature in the midst of production and human livelihoods. Thus, concurring with the thoughts of Marx (1988), with the emergence of industry the Earth becomes a whole arsenal of natural resources, a place where nature and people converge into factors of production. Consequently, the factors of production converge on a system that organises surroundings for the transformation of these factors into products. This transformation is accomplished through work. It could be affirmed that the industrial revolution and the consequent development of technology generated and promoted new concepts of social organisation, of social movements and the appropriation of nature.

As Soares (2008) has argued, the industrial revolution brought with it human-worker training. However, processing time and skills, coupled with the exponential growth of industries, resulted in changed relationships since workers were faced with the need for technical expertise to operate industrial machinery. As well, workers, using the relationship between themselves and machines, consolidated the worker-power-of-work concept. The worker-power-of-work concept is oriented and organised by the time-production process.

The home-work-force relationship becomes a reference point for the procurement of goods and services within the capitalist system. The worker's productive strength delineates the goods to which he or she has access. Thus, the value of an individual's work becomes the value of his or her production and consequently his or her potential. Following this logic, the worker-power-of-work becomes the human-factor-of-production. Supported by the consolidation of capitalism as a hegemonic system, the human factor-of-production will be recognised by the company not for the workers' productivity but by their ability to consume. Thus, consumption becomes exponentially exacerbated, transforming the human-factor-of-production into the human-factor-of-use. The human-factor-consumption has its social recognition in the ability to consume goods and services. In this way nature is exploited and consumption is raw and unbridled. The *modus vivendi* in society is a cycle fuelled by a cash return resulting from the production cycle.

Within the cycle of production and consumption of goods and services there is also a strong concern with the exploitation and increasing artificiality of nature. However, even today it is possible to realise that many times this concern does not come accompanied by an understanding of the ways in which natural resources are formed and, where possible, renewed. There is a problem of establishing a balance between people's needs and finite natural resources.

The balance between the growth of consumption of natural resources and the growth of the population that consumes them engenders in human-factor-consumption the understanding that consumption is essential to the survival of society. However, this internalisation will come accompanied by an understanding that people are responsible for the preservation and maintenance of these same features. We are dealing with the person-consumer, an essential component in the population for the treatment of socio-environmental issues. The person-consumer is part of the balance in the proportions of the use of natural resources and the problems of food. As well, the person-consumer becomes critical to the future as a forecaster of the possibilities for the production, storage and, mainly, the exploitation of natural resources. These forecasts are usually the product of census results and the plans for government action or the results of human interactions of consumers with their surroundings.

As a source of data for future forecasts, the person-consumer within economic policies and the re-adaptation of the production and use of nature becomes a person-statistician This is a person responsible for the establishment of the current dynamics of production and consumption within the future expectations of society in which he or she lives and the society which he or she seeks to form. The person-statistician is concerned about the ways in which nature is explored and exploited and searches, via practical activities, to find opportunities for reframing the dynamics of environmental exploration. However, the person-statistician has another challenge ahead: to work to change the momentum established from the beginning of the industrial revolution.

Understanding the dynamics of human evolution can bring about changes in the relationships between people and nature or between people and changes in the labour situation. These modifications occur according to the wishes and needs of the society in which people belong. Insofar as these relationships change, society changes instantly. There are changes in social behaviour that will determine how a child, for example, matures and grows old as well as when the child becomes old.

These clarifications are important to help us visualise how humans change as education guides their behaviour towards the dominant interests (education for the labour market and the acquisition of productive skills) and how the interests of a certain society 'shape' human activity. Thus, being educated means finding out that "the human being is essentially a person who has basic needs such as, for example, eating (a hungry animal), whose needs must be met and, therefore, there must be consumption" (Boff 2006: 35). We also know that human beings assured their own consumption without being concerned for the needs of their heirs, that is, without the notion of sustainability.

The idea of building a sustainable society, by following the guidelines outlined above, is an idea that gains dimension and concreteness. This way, a sustainable society requires 'environmentally educated citizens' and by this we mean citizens who are able as individuals to participate in the social transformation of the world in which they live or, as Mazochi and Carvalho (2008) contend, individuals who are capable of,

1. understanding the environmental problems of the current world (the concepts and information dimension);
2. understanding oneself personally while being in social groups in relation to this problematic (the size value);
3. having the capability to act effectively on these matters (the political participation dimension).

With regard to the project *My Neighbourhood, My Land, My Treasure*, that we shall discuss later in this chapter, it is claimed that through the clarity of understanding we are educated to believe that we are 'beings', but we are also educated to understand the causes and consequences arising out of consumption. In other words, the current education system develops the potential to work, but disregards the development of citizenship. The discussion we propose relates to the means by which we can act against exaggerated consumption and advanced production. We propose to discuss human relationships and the autonomy of individuals faced by contemporary conditions – globalisation of the economy and production systems, fragmentation of individuals and of the social order, a genetic and informational revolution, among others.

School, Society and Sustainable Development: The Case of the Paulo Tapajós College, São Paulo/Brazil

In a publication of the Brazilian Ministry of Education titled *Education for Citizenship. Education Guide for Sustainability: Earth Charter* (December 2006) the concept of sustainable development has as its essence the following premise: 'Enough for everyone, forever'. This concept is paramount for humans if education is to respect the common good. This implies reflection on the care of all existing resources on Planet Earth in order to provide a good quality of life for all generations. This educational commitment can be interpreted as an education that leads people to have a "universal responsibility" (Ministry of Education 2006: 18), i.e., the recognition that "we belong to the land; we are sons and daughters of the Earth; we are the Earth" (Boff 2006: 33). Today, in the case of an education for future generations, reflection on practice from this perspective cannot be seen as something missionary but as an action for social organisation which in turn focuses on the use of resources. Any social action will have consequences visible in the future.

Combined with the idea of action with a view to the future there is the understanding that all human creation serves to satisfy the needs of people. "Insofar as man creates, for example, technologies within an historic process, these same technologies are used as the basis for cultural advancement" (Petarnella 2008: 45). Building on this idea, we realise how actions in the present threaten the future of humanity.

Returning to the principles referred to at the beginning of this chapter, the school, especially the teaching and learning environment, provides opportunities for reflecting on the environment and a sustainable lifestyle. Social changes in schools will be brought about through social agents, by people. By understanding the importance of now as a reflection for tomorrow, "it is fair to bring aid to those who will follow" (Pessanha 2007: 79), given that the sustainable activities of today are perpetuated through socio-historical processes of which people are a part.

Values assigned to the environment and sustainability must be grounded at the heart of human culture. When these values are taken as ethical principles (and the school plays an extremely important role in this process), these same values become part of human nature, allowing people to project a future from these ethical principles. This way one can perform, with minimal risk, future forecasting and designing this in a conscious and responsible way.

Understanding to ensure the future is an issue that involves individual values that must go beyond discursive activities. It is worth remembering that in a not so distant past, under the ideal of human emancipation through the acceleration of progress, people possessed the production process, the environment and all resources that are in it. As we saw at the beginning of this chapter, this belief became even more rooted in the first industrial revolution. This occurred through the creation of the idea that a person could be an individual "competitive consumer, essentially someone who could quickly waste what was produced, given the disposable value of things..." (Lima 2009). However, this idea did not take into account the fact that human survival depended on natural resources and most of these were exhaustible.

With the industrial revolution in Brazil, the education system was geared to meet the needs of production and industry. Changes in the education system and the expansion of primary schools in the early twentieth century contributed to students not reflecting on the fact that their survival was directly linked to their capacity to engage in their own actions. Until the early 1970s in Brazil these activities were totally dedicated to production and consequently to the consumption of resources and with a disregard for the possibilities of development and progress in a sustainable way. For this reason, schools today have invested in the idea that, "There is a need to promote through education the values, behaviour and lifestyles that are essential to a sustainable future" (Ministry of Education 2006: 15).

By following the guidelines of the Ministry of Education with respect to the introduction of the environment as a theme to be worked on and within schools the school takes on a political and ideological role, becoming the locus for conscientizing and forming ethical citizens. To do so, schools must include in their plans creative methods for student development through reflective practice. This means providing, in addition to knowledge and information included in the curriculum subjects, the bases for qualitative activities whose results are directed at "personal satisfaction in the present without compromising the future" (Ministry of Education 2006: 18), i.e., designing for today while designing for the future.

To understand better the intervention of the school as an educative force in society, one has to think about the disciplinary content. Considerations of this content should go beyond simply an annotation on a sheet of paper or any other resource. The disciplinary content must go beyond mere statements as commonly viewed in Brazilian schools. Also, because the content viewed in this way serves only to meet bureaucratic needs, providing a framework for teaching, it does not guarantee the student's learning. Often, this content is merely a quantitative index in government statistics.

Schooling, when thought of as a body of content to be acquired from disciplines, collaborates with a view of education as training focused on the transmission of information in a non-critical way that seeks to strengthen the understanding of society as a "major consumption centre". This view of training encompasses the production of knowledge oriented towards the unbridled use of non-renewable resources and the performance of actions that repeat and ensure the order of industrialisation. In contrast to this view, that emphasises progress and profit at any price, there are possibilities of perceiving curriculum content in schools as a way of arousing awareness. Such awareness includes a consideration of the effects of today's behaviour on tomorrow's world and an understanding that an educator has the capacity to engage in education as essentially a liberating activity in which learning from the past helps to build a better future.

Understanding education as a way of promoting critical awareness, educators seek to encourage awareness practices that result in the rooting of sustainable activities in learners. Through the process of teaching, critically aware and environmentally conscious educators seek to transform students into social agents. In this context, the disciplinary content must "be participatory and inclusive, be interdisciplinary, be student-centred, see students as subjects who simultaneously learn and teach, contextualised and be flexible, be collaborative, be transformative" (Ministry of Education 2006: 16).

With regard to education for sustainability, this form of education provides a basis for the political and ideological positioning of schools. Schools must devise ways to develop the potential of today's world, advancing on educational models that expose students to reliable data and inviting them to be reflective practitioners. In this sense, the school's role is to develop an attitude in learners of 'taking care of themselves' as a means of human emancipation and 'caring for themselves' as a way of balancing and maintaining mankind and the environment.

My Neighbourhood, My Land, My Treasure

Based on the principles set out above, the project *My Neighbourhood, My Land, My Treasure*, was presented in 2007 to the Board of Education of Mogi das Cruzes/SP – the representative body of the Secretariat of Education of the State of São Paulo in the city. The project was undertaken in Paulo Tapajós College, an elementary school. Initially the project sought to integrate the school community and the local community to engage in activities aimed at the sustainable development of the places where students lived as social actors. The project started from the understanding that the school could contribute to the creation of a community with macro participation. According to Bordenave (1987: 24–25) such macro participation,

> ...includes the involvement of people in the dynamic processes that are or have changed the society. Its conceptualisation, therefore, should focus on what is most basic in society, which is the production of material and culture, as well as its administration and its fruition. Under this premise, social participation is the process whereby the various social layers have a part in the production, management and the enjoyment of goods of a society historically determined.

In the project, the participants were children 8–12 years old and their families lived in a rural area at the foot of the Atlantic forest in Mogi das Cruzes/SP. *The My Neighbourhood, My Land, My Treasure* project was developed and coordinated by Professor Eduardo Campos Garcia. It prioritised the critical and reflective education of children so that these children could contribute to transforming their community. The importance of such training fits in with the structuring of the "Brazilian environmental movement, which assumes a multi-sector configuration requiring actors with practices focused on finding a viable conservation" (Jacobi 1999: 177) of the environment.

In the project proposal, the change agent is not a person who simply receives certain information and acts within the content limits set by the curriculum. The change agent in the project proposal is a person who seeks by means of knowledge acquired in the community and at school to address questions about his or her future and the future of other generations in a school subject that created connections between the learning process, engagement, awareness, policy and habitat. The *My Neighbourhood, My Land, My Treasure* project sought the formation of a student citizen, a constructive agent who is conscious of the present and future demands of his or her social surroundings.

Achieving the objectives proposed by the project was not an easy task. In a predominantly agricultural region the daily need for family survival took priority over unknown aspects of the future. The family members of the students were sharecroppers whose relationships with the landowners were reminiscent of the feudal period. Since these students were the children of sharecroppers, problems related to labour issues undermined the momentum, including the possibilities of transformation of the students participating in the project.

For community awareness in *My Neighbourhood, My Land, My Treasure* the starting point was a description of the belongings and the surroundings of the community.

The 160 students involved in the project engaged in fieldwork in order to collect information about what the residents of the community regarded as significant. In this initial stage, we did not want to provide information about situations and conditions. We wanted these students to be aware of what underscored their own practices.

Given the need to understand the problems in their community, students lacked a method or an academic rigour to achieve this purpose. We knew that the problems about which the students were thinking did not have any ready-made or immediate solutions. The initial proposal was to promote a business in the community, thus breaking out of the four walls of the classroom. The importance of breaking out of the four walls of the classroom to have an educational practice in the community is suggested by Begatini and Bertelli (2005: 2) because,

> With the strengthening of the environmental programme in schools and in the mobilisation of children, it is believed that education, through a pedagogical dimension that encourages leaders in training, you will find a new path, where the relationships between man and nature at the local, regional and national levels become relevant aspects of the preservation of the biota association as a source of life for the planet.

Through students' studies of regional biota, it was essential for the school to become a place for reflection on these issues. In this way, the project was driven to collect information about the events that occurred in the neighbourhood in which the school was located. Professor Eduardo Campos was the person responsible for the school-student-community articulation. Students presented their initial reports and were encouraged to seek causes in a preliminary way. We agreed with Soares (2009: 13) that this process was fundamental because "…these thoughts lead us to understand why school education is a process in which the teacher and his students relate to the world through relations that hinder each other at school and in ideas".

At the time of the project, devastation occurred in the daily life of the rural district of Manoel Ferreira when the school burned down. In addition to this, the community had numerous environmental problems, such as trafficking in animals, poor sanitation, and the contamination of the riparian forest. From students' reports, the teacher asked the children about their everyday life as expressed in drawings, as well as elucidating their thinking about the answers written in their reports. The drawings showed that these students had conflicts between the content learned in school and their experiences outside. Conflicts emerged because the burning of the forest for economic purposes, for example, was something 'naturally accepted' by the community. Fires occurred outside local procedures and were accepted by the bodies responsible for the preservation of the environment. In contrast these fires maintained at the same time many families including the families of these students involved in the project.

The second stage of the project, was questioning the students to find out what their drawings represented to them. This quest became local news. When referring to a drawing by a student, on September 16, 2006, *Newszinho* (Mogi News newspaper supplement: 8) reported: "These drawings show a sad moment, where the student

J.C. portrays a great fire in a nearby school…students were disappointed with the scenes of destruction …". In fact, what we saw was nature turning into ashes from the burning of an area that should be preserved and this fact has been repeated numerous times with the same characteristics. The fire scene was a scene that saddened and disappointed all of the students and their teachers.

A repercussion of this was the arrangement of discussions with students about their living environment. From these discussions, we were able to show the need to develop further activities within the project – activities that placed a high value on their lives. Thus, students were invited to 'wander' through their surroundings, pointing out things and circumstances that could in one way or another inhibit the development of life. These students became immersed in defending the environment and hence their disappointment with the actions of the community grew as the project developed.

Through the walk, we were able to list a series of features that tarnished daily life. These features prevented what we would come to call sustainable living. Students referred to the use and disposal of plastic packaging of agrochemicals in the middle of the forest. The residues from pesticides contaminated the springs that surrounded the community. The infection of streams by animal and human faeces, as well as deforestation, caused considerable damage to Brazilian flora and fauna. The sale of bio-native species was another of the problems perceived and raised by students.

From this point, the awareness of students became political engagement. We shared the ideas proposed by Leo Maar who argued that, "we must think about the past that hurt the present for this injured present could not happen in future" (2006: 11). Consequently, school actions emerged directed exclusively at the environment and they emphasised the engagement of students in a sustainable manner in their community. So, these same students started to develop projects in their neighbourhood to educate the local population about the environment and in this the project was successful.

Autonomy was the key to the project and reflection the starting point for all activities. Thus, in questioning the past, seeking evidence and proposing practical solutions for the future, students began to seek ways to suspend their daily life and promote attitudes that could bring a return in the short, medium and long terms.

Results of the *My Neighbourhood, My Land, My Treasure* Project

The breaking down of the barrier between the concrete and the abstract is started in school. Reflexive classes were literature based with students reading *A Midsummer Night's Dream* by Shakespeare as required in the Law of Guidelines and Bases of Education – LDB (1996). Article 28 i.b.3.2.4 posits the importance of education for citizenship and autonomy. Thus, "The curriculum content and methodologies should be appropriate to the real needs and interests of students in the rural area", among them the knowledge about the inhabitants and the physical aspects of the neighbourhood. It is important to notice Article 32 that refers to education as,

> Elementary, with a minimum duration of eight years, compulsory and free in the public school, must aim at the basic training of the citizen, by: I – developing the capacity to complete mastery of the basic reading…; II – understanding of the natural and social environment, political system, technology, arts and values on which society is based; III – development of learning capacity, with a view to acquiring knowledge and skills and the formation of attitudes and values; IV – strengthening family bonds, bonds of human solidarity and mutual tolerance in which sits the social life.

According to the National Curriculum Parameters – PCNs, "The exercise of citizenship requires access to all the cultural resources relevant to intervention and responsible participation in social life" (1997: 27). Reading *A Midsummer Night's Dream* was intended to promote consideration by students of abstract and concrete possibilities. For 8 year old pupils, the mediator was essentially saying that the attitude of some landowners was 'evil'. These interpretations drew on memories of landowners as persons who controlled their plantations and their employees. On occasion, these employees were relatives of the students.

On the one hand, there was pressure to tell students what they needed to know and, on the other hand, there was a need to guide students as to how they should undertake their studies. The first external activity focused on the neighbourhood water supply. The district had three water sources where human intervention was obvious. Secondly, to evaluate the ignorance of the population with regard to sustainability, a pipe and a water reservoir were put on site. In a spring, some houses had been dumping oil that had been mixed with human waste and the residues of pesticides. Students also revealed the existence of a clandestine garbage dump in the middle of the forest and this had been used to discard the packaging of agrochemicals. This dump was unknown to many of the community residents.

Mediated by their teacher, students came together and decided they would mount a campaign focused on the importance of preserving water. More than that, they wished to emphasise the importance of the community caring for existing neighbourhood water sources as something important for future generations. This activity has echoes of Boff's (2006: 111) position that "there was a struggle for survival of the fittest which ensured the persistence of life and individuals until today, but co-operation and co-existence among them" motivated the actions of the students.

Students designed and distributed posters with the theme "Using water responsibly, its re-use and preservation…". The posters were distributed to the neighbourhood shops. At the time of delivery, students chatted with shopowners about the use they made of water. In these conversations, they discovered that the water was never reused, even as water for cleaning sidewalks. On these findings, the students began to draw up plans for community awareness, questioning the real reasons that led villagers to waste natural resources. In public meetings arranged at school, students showed the importance of taking care of the well water so that future generations could enjoy the same quality of water as enjoyed by the current residents of the neighbourhood. At these meetings, the students asked those present about the disposal of chemical waste plastic packaging (agrochemicals) in the neighbourhood forest and around watersheds. The students' posed hard-hitting questions about this disposition and the reasons for the discards by those farmers who were not concerned about the potential scarcity of natural resources.

Subsequently, in a class activity that simulated a lawsuit the students wrote a letter that they sent to the Board of Education of Mogi das Cruzes, the body responsible for managing and monitoring school activities. The Board in turn triggered the rural syndicate regarding the claims made by the students. A request was made to the rural syndicate to provide appropriate informational material so that students could clarify the correct way of handling and disposing of residual pesticides as garbage and know about the care and protection needed relating to their handling.

These small-scale activities of students from School Paul Tapajós, located in the municipality of Mogi das Cruzes began to produce results. The Union of Rural Workers provided a workshop in the school for the community. The syndicate also provided a manual *School Project in the Field: Signet* that provided information about the care required in the handling, use and disposal of packaging of agrochemicals. Intervention by the Board of Education and the rural community syndicate brought changes in the behaviour of this community with regard to nature and natural resources.

Students created a magazine at school that was intended to offer instruction about caring for the environment as well as educating the community about it. In the magazine, students treated the importance of responsible care of forests, lands and awareness about the use of water, disposal of chemical waste materials, the re-use of vegetable oil and issues related to animals. Simultaneously in school, students engaged in studies including such concepts as aquifer, extinction, destruction, and contamination, with which they were familiar in their everyday lives. From their reflection on their experiences of life, students collectively produced a theatrical play entitled "I'm not the owner of the world, but I'm a son of Donna". In this work, the text-turned-theatre piece addressed the care that was needed for the planet. The Earth was understood to be the mother of all humans who were at the mercy of fate. This fate began to change when a lumberjack decided to talk to a tree that, in turn, gave an account of its sadness. The greatest sadness of the tree was concerned with the future of the lumberjack who had not realised how much he harmed himself. The Board of Directors of Teaching in Mogi das Cruzes arranged for this piece to be presented to 500 students from the public school network and to the local community.

In the period in which the project was developed, the Mayor of the city of Mogi das Cruzes initiated a debate about locating a garbage dump in the neighbourhood of Helens, an industrial district of the municipality. The local press reported discussions on the future installation of a large dump. The students began to read and interact with the journalistic genre realising that the problems of non-sustainable development also served the interests of entrepreneurs. In Brazil, financial and political agreements are constantly concluded without taking sustainability into account. In such agreements, references are made to sustainability awareness and nature simply as rhetoric and not as aspects of effective actions.

Following their reading about the problems highlighted by the press that week, students, through a newsletter, started to consider the problems caused by garbage in the neighbourhood. On account of their campaign and the political controversy stirred up by these students, the rural syndicate gave each student an identity card as an environmental agent. Through this identification, the students began to position themselves in front of the local residents in relation to the preservation of their natural resources. These students had become agents with a share in the community in which they lived.

At the time of writing, the project has not been finalised. The continuing activity of this project has been the engagement of community residents and students in community groups and the conscious recognition of the ethical and sustainable actions of these students.

Reflecting on the Project

From the second industrial revolution, production and food consumption were dreams. Dreams, as Barbrook (2009: 15) reminds us, are, "anticipating the future, because man has the objective to realise your dreams". This form of designing for the future originates in a, "second industrialisation that becomes the industrialisation of the spirit, and a second colonisation that becomes the colonisation of the soul". While the first industrial revolution sought to transform the human factor of production – humans became part of a product to be consumed – the second industrial revolution sought to industrialise the way of thinking and acting of the human being, stressing what came to be called mass culture.

In seeking to understand the needs of the imagined future, it is necessary to consider the possibilities for actions leading to the development of a policy to promote citizens' awareness. There is an opportunity here for breaking down some of the current dominant thinking and the schools have a crucial role in this. For everyday exercises in schools, we propose including a view of the future in which students might participate that takes into account possible causes, possible actions and, mainly, possible solutions for what could be considered a disaster for future humanity. In these exercises, it should be considered that,

> Nature is at the same time organic and inorganic, fragmentary and unitary, mechanical and alive. It implies a synthesis that is not the "sum of all the parts" but feedback, transfiguration, recombination, categories of movement that take nature as it constantly unifies and diversifies. From the point of view of the synthesis of life, we can infer the transfiguration of inorganic flora in the process of photosynthesis in fauna and flora and the whole organic sphere turning on human existence (Soares 2009: 15).

So as not to suffer the consequences that could be disastrous for the future, people in society must be rethought of as beings of nature and at the same time products of history. This way the people become agents responsible for processing, constructing and reconstructing society. While being natural, living in the biotic sphere, people are structured and organised as animals with the gift of reason who seek to perpetuate the human species, improving it through continuous evolution. However, the structuring of society around the needs of capital places people in a natural condition of human existence. Whereas the humanisation of man and nature is the third step in the development of the human spirit and ideas, this is the current legacy left by the first and the second industrial revolutions. The schools transmit the legacy of these revolutions to students. With regard to schools, it is considered that at a given moment in cultural history education has been concerned with rampant consumerism and the prevalence of capitalist production, where it was necessary to produce more in less time, making the worker a "cog in the machine" (Chaplin 1936, Chapter I). Today it is argued that education for sustainability should be founded on five pillars: "social, economic, ecological, cultural and spatial sustainability" (Jacobi 1999: 176). Building on these five pillars, the school engages in the formation of citizens with a broad vision of the world. As a result, "the clarity of the teaching act is a political and technical expertise as it relates to life, the history of man and nature as well as yourself" (Soares 2009: 15). Thus, the values represented in the whole citizen are nothing more than an understanding of people and nature as common goods with interdependent common interests.

Since people are nature and nature humanises people, there is an interest in this relationship that should go beyond speech. This relationship should lead people to think about their actions because without the *praxis* development of citizens, local actions do not translate into actions for sustainability. There is a failure to generate the ideas needed about the global consequences of those same actions. This implies that an education for sustainable development is for both today and tomorrow and is a matter of ethical and moral development and not just isolated actions.

A school that does not pay attention to social questions and to social problems does not contribute to the formation of thought that meets the needs of students for life, at any time and in any place. As a result, there is a need to reformulate the values of education. This involves breaking away from the dominant paradigms that emphasise the role of education as a way to production and shifting towards education as being concerned with the quality of life now and in the future. Within this proposal, educators adopt the role not as cogs in a machine but as agents who participate in the planning process so that the results are more edifying.

The school is a fundamental institution for breaking down those technological paradigms that occur at the expense of strengthening the autonomy of the subject and the development of those ethical and moral principles associated with sustainability. For schools, activities like those developed in the *My Neighbourhood, My Land, My Treasure* project are fundamental to the notion of sustainable development, thus, "society is transformed, creating a major change in their own civilising process" (Jacobi 1999: 178).

With regard to teachers, these activities will provide the *praxis* only if they are driven by an emancipatory vision that is geared towards building learning communities. These communities would promote the development of the capacity for reflection, innovation, flexibility and control of the problems created by capitalist society. Teaching has to overcome the distance between theories, ideals and daily practices, including excessive consumption, loss of a sense of community and the growing estrangement between social classes.

In taking responsibility for education to promote sustainable thinking, the school engages in political activity. In this sense the *My Neighbourhood, My Land, My Treasure* project fulfils its role and its intention: to promote an awareness on the part of those involved, that, "the human being is both natural and history-temporal" (Boff 2006: 67), and an awareness that, "all conduct implies modification to the construction of final values" (Piaget 1978: 21). The results of this have important consequences for the future.

Conclusion

Today, reflecting critically on autonomy, about the identity of individuals and their roles in social interactions, is reflecting on the fact that we are people who are naturally human and humanly natural. This means that we are part of nature before being a product or output. To reflect on nature is to think deeply about our own lives and to realise our responsibility to protect and conserve the natural environment. This reflexive activity, thinking about the future while conserving aspects of the past, is a function of schools acting in their local communities. Schools have the potential to become transformative agents for the environments in which they operate. Their importance as transformative agents is closely linked with interventions that promote changes in society and in the local environment. This implies that the school could be seen 'as a person' knowingly interfering in society, taking responsibility for the development of activities that contribute to an understanding of the environmental framework of which it is a part.

To ensure that they contribute to education for sustainable development, schools should align formal education with citizenship education. This alignment can occur by changing the curriculum to allow a greater integration between the school and its community. An example of this is the work of the *My Neighbourhood, My Land, My Treasure* project, where the school walls no longer limit the space for teaching and learning. Activities such as those developed by the *My Neighbourhood, My Land, My Treasure* project faced difficulties regarding their acceptance and implementation in public schools, but the purposes of the activities are justified in the use of sustainable thinking as a guideline for building an identity based on solidarity and the generation of autonomy.

We end this chapter mindful that those who talk, teach and lecture about sustainability should themselves behave sustainably. The implementation of education for sustainability projects becomes an arduous task when these activities go against the logic dominant since the industrial revolution, a logic focused on the production and consumption of goods. In contrast, we know that to educate for sustainability is to educate for autonomy, is to educate for the awareness that we are all responsible for everything around us, and around us is the land from which all life springs.

References

Barbrook, R. 2009. *Futuros Imaginários: das Máquinas Pensantes à Aldeia Global*. Rio de Janeiro: Editora Petrópolis.

Begatini, R. M., and Bertelli, M. D. 2005. Educação Ambiental: inter e transdisciplinariedade – Tema abordado em uma escola rururbana). *IV Encontro Ibero Americano de Coletivos escolares e redes de professores que fazem investigação na sua escola.*

Boff, L. 2006. *Saber. Virtudes para um Mundo Possível: Hospitalidade*. Petrópolis: Vozes.

Bordenave, J. E. D. 1987. *O que é participação*. São Paulo: Brasiliense.

Chaplin, C. (Director) 1936. *Modern times* (Filme Cinematográfico). EUA: Continental.

Jacobi, P. 1999. Meio Ambiente e sustentabilidade: O complexo desafio da sustentabilidade. In *O Município no Século XXI: Cenários e Perspectivas*, ed. P Jacobi. São Paulo: CEPAM.

Leo Maar, W. 2006. À Guisa de Introdução: Adorno e a Experiência Formativa. In *Educação e Emancipação*, ed. T W Adorno. Rio de Janeiro: Paz e Terra.

Lima, G. B. 2009. *A Abrangência Histórica da Revolução Industrial e Seus Desdobramentossociais, Econômicos, e Ambientais: Uma Análise Contemporânea*. http://www.cenariointernacional.com.br/artigos2.asp?id=131. Accessed 28 February 2010.

Marx, K. 1988. *O Capital: Crítica da Economia Política*. Vol. II, Book. 2, *O Processo de Produção do Capital*. São Paulo: Nova cultural.

Mazochi, L. M. and Carvalho L. H. 2008. *Educação Ambiental Formadora de Cidadania em Perspectiva Emancipatória: Constituição de Uma Proposta Para Formação Continuada de Professores*. XXXI Annual Meeting of the Association of Graduate Education – ANPED, Caxambú (XXXI Reunião Anual da Associação de Pós-Graduação em Educação – ANPED/ Caxambú): http://www.anped.org.br/reunioes/31ra/1trabalho/GT22-4903--Int.pdf. Accessed: 18 June 2010.

Ministério da Educação. 1997. *Parâmetros Curriculares Nacionais*. Brasília: MEC/SEF.

Ministério da Educação. 2006. *Educação Para a Cidadania. Guião de Educação Para a Sustentabilidade: Carta da Terra*. Brasília: Ministério da Educação.

Morin, E. 1969. *Cultura de Massas no Século XXI: O Espírito do Tempo*. Rio de Janeiro: Forense.

Pessanha, J. A. 2007. As delícias do Jardim. In *Ética: Vários Autores*, ed. A Novais. São Paulo: Companhia de Bolso.

Petarnella, L. 2008. *Escola Analógica, Cabeças Digitais: O Cotidiano Escolar Frente às Tecnologias Midiáticas e Digitais de Informação e Comunicação*. Campinas: Alínea.

Piaget, J. 1978. *Seis Estudos de Psicologia*. Rio de Janeiro: Forense.

Santos, M. 1994. *Técnica Espaço Tempo*. São Paulo: Hucitec.

Soares, M. L. 2008. *Da evolução da Concepção de Natureza e de Homem na Ambiência de Uma Educação Ambiental Crítica*. XXXI Reunião Anual da Associação de Pós-Graduação em Educação – ANPED/Caxambú: http://www.anped.org.br/reunioes/31ra/1trabalho/GT22-4153--Int.pdf. Accessed 10 June 2010.

Soares, M. L. 2009 *Entre a Dupla Determinação de Homem e a Revolução Técnico-Científica no Campo da Educação Ambiental Crítica*. XXXII Reunião Anual da Associação de Pós-Graduação em Educação – ANPED/Caxambú: http://www.anped.org.br/reunioes/32ra/arquivos/trabalhos/GT22-5165--Int.pdf. Accessed 10 June 2010.

Chapter 10
Higher Education and Regional Sustainable Development: The Case of the Federal University of Tocantins in the Brazilian Cerrado

Carmen Lucia Artioli Rolim, José Damião Trindade Rocha, and Paulo Alexandre Adler Pereira

Introduction

This chapter focuses on the creation of the Federal University of Tocantins (UFT) in the Legal Brazilian Amazon region. We present the main lines of a systemic vision related to the subject as a strategy for locating readers in the big picture and motivating new approaches to problems and questions intrinsic to the object itself. First of all, we must ask: Why is the creation of the UFT regarded as a unique event in the national and international scene?

The UFT was created in 2003 with a structure of seven campuses in the state of Tocantins with its headquarters in the city of Palmas. The UFT is consequently located in the Legal Brazilian Amazon region, a geopolitical division created in 1953, reaffirmed in 1966 and consolidated by recent public policy. Eventually, the state of Tocantins was formally created in 1988 through Article 13 of the Act of Provisions for Transitory Constitutions of the Federal Republic Constitution of Brazil. Soon after, the city of Miracema do Tocantins was chosen as the temporary capital of Tocantins and the construction of the new capital 'Palmas' was initiated. In a short time, that is, in 1989, it was inaugurated as the most recent planned city of the twentieth century.

The state of Tocantins is emblematic in many aspects. Although it is located in the Amazon region, its territory lies in the transition zone of major Brazilian ecosystems: the Amazon forest, the *cerrado* (tropical savanna), the semi-arid and the *pantanal* (wetlands). It has rich biological and cultural diversity together with abundant natural resources. Looking closely, the division of the Amazon is quite recent and represents an increasingly problematic and significant focus of national and international

C.L.A. Rolim (✉) • J.D.T. Rocha • P.A.A. Pereira
Federal University of Tocantins, Palmas, Tocantins, Brazil
e-mail: carmem.rolim@mail.uft.edu.br

M.L. de Amorim Soares and L. Petarnella (eds.), *Schooling for Sustainable Development in South America*, Schooling for Sustainable Development 2,
DOI 10.1007/978-94-007-1754-1_10,

attention, representing a typical scenario for discussions on sustainability. A new state and its planned capital were created over 20 years in the Amazon. It is a multicultural capital built and occupied by citizens from different municipalities and states in Brazil. In the meantime, in 2003 the UFT was created with campuses located in seven counties in the state of Tocantins forming a north-south axis that connects the edges of the state. The location of the campuses at distances ranging between 70 and 600 km from the capital poses a number of challenges.

Its commitment to sustainable development of the Amazon appears in two of its major institutional planning documents: the Institutional Development Plan (PDI, Plano de Desenvolvimento Institucional); and the Institutional-Educational Project (PPI, Projeto Pedagógico-Institucional).

> UFT Mission: The Mission of the University of Tocantins is to produce and disseminate knowledge to educate citizens and qualified professionals committed to sustainable development in Amazonia…
>
> Strategic vision of UFT: By 2010, the UFT will be a consolidated University, multi-campuses, a democratic space for cultural expression, recognised for its high quality education and research and extension activities aimed at sustainable development in the Amazon (University of Tocantins 2003: 8).

In 21 years Palmas left the drawing boards of engineers and became a structured and consolidated city. In 7 years, the UFT has consolidated its institutional progress, structuring its organisational planning, developing a balance of teaching, research and extension activities, and has currently 43 undergraduate courses, seven masters courses, one doctoral course, three doctoral courses in partnership with other institutions and about 567 permanent teachers including 224 Doctors, 316 Masters and 27 experts, in addition to 400 employees and 10,000 students. It has instituted democratic management, held elections for its leaders, having its first president elected on August 20, 2003.

The creation of the UFT has produced a considerable local and regional impact. Initially, it is possible to consider this impact using the concept of externality. It is common to assess the impact of an economic enterprise through the externalities it creates. In general, the externality can suggest how the effects of decisions or ventures unrelated to their own community are shared by members of a particular community.

Externalities can be positive or negative, depending on whether they lead to benefits or losses to the whole population. In principle, there is an agreement under which public services in Brazil including education and scientific research always produce positive externalities,

> Education generates positive externalities because the members of a society and not only the students receive the many benefits arising from the existence of a more educated population and are not accounted by the market. Thus, for example, several studies based on different methodologies show that education helps improve the levels of health of a given population. Especially, higher levels of maternal education reduce child mortality rates. Other studies also show that education helps to reduce crime. All these indirect benefits of education because they are not priced are not computed in the private benefits. Therefore, the social benefits outweigh the private benefits, which include only the personal benefits of education, for example, wages obtained depending on the level of schooling (Souza 2008: 2).

Based on these considerations, could we be satisfied with the thesis that the creation of the UFT has produced a set of positive externalities far-reaching in the Amazon region and Tocantins state? We believe that the problem transcends the perspective guided by the economic costs and benefits of decisions in today's society.

A relevant point of view about the creation of the UFT is the territorial dialectics which forms the complex scenario of the Legal Amazon. This scenario can be represented by the following pattern:

- natural territory: qualified by the establishment of conservation units;
- indigenous territory: qualified by the integration between biodiversity and local cultural diversity;
- urban territories: qualified by spatial adjustments, administrative divisions and the legal system;
- adapted rural areas: qualified by economic production with a predominance of farming.

Such territorial divisions tend to form spaces classified by the types of human undertaking, their inter-relationships and projects. They define a scenario in which the spaces are components of environmental issues that challenge the Amazon and in different ways attract the attention of international society. This scenario is endowed with a high dose of abstraction. But the simplified model allows for tracking the creation of UFT as an area described by the training activities of various segments of the population through a greater commitment with respect to the sustained and sustainable development of Amazonia. This poses a great challenge for the UFT.

Contextualising Development in the Amazon

The Amazon was the subject of various development policies that from one extreme to another aimed at the geopolitical occupation of the territory in line with economic development, mainly in the perspective of modernisation at all costs. Interventions caused by large public works, such as hydroelectric plants and highways, occurred with no regard to natural, indigenous, urban and rural areas, often producing negative environmental impacts and environmental conflicts with disastrous consequences. Amazonian biodiversity remained virtually intact until the 1970s when the Trans-Amazon highway was inaugurated. This was the starting point for the 'modern era' of deforestation. There was also a significant increase in deforestation rates since 1991 with the expansion of agricultural activities as a major component.

The gradual adoption of the environmental component in public policies at the national level indicates progress towards sustainable practices. However, conflicts and contradictions observed in a significant way since the 1970s are still visible with the operations of public works undertaken by the Programme to Accelerate the Growth of the Federal Government (PAC, Programa de Aceleração do Crescimento do Governo Federal). An example of this can be seen in the relationship between the actions of the PAC and the vision of indigenous people regarding environmental issues and potential impacts of PAC on their land. From this assumption we continue to live with a series of threats to the sustainability of the Amazon that require the consideration of actions that produce positive impacts on the region. In fact, we consider that educational actions should be prioritised, allowing the formation of new generations of citizens who need to be prepared to deal with the various dimensions that are involved in social and environmental problems.

It is increasingly clear that global society faces a major civilisation challenge with respect to environmental problems and development models. Environmental issues are progressively filling national and international agendas no longer as collateral challenges but as systemic problems. The latter years of the twentieth century were marked by changes beyond the socio-economic field, involving political and cultural aspects. For Carril (2007), the United Nations Conference on the Environment has led to worldwide concerns over natural resources affected by predatory economic actions. In the words of Vieira (1997), "...theories of economic development of the twentieth century, and the resulting economic policies, always ignored the environmental condition, considered only as an externality" (p.126). For Schramm (1999) the twentieth century has seen substantial advances in technology but has shown the largest strain on the environment. With Carril (2007: 62) we see that,

> ...historically..., the capitalism subsidised by science and modern technology has consolidated the process of dehumanisation of nature and denaturising of the human being, prepared by the steps of the modern science construction based on rationalism. This approach confirmed the reciprocal externalities between humans and nature, that is, the human being understood as being excluded from the concept of nature, standing up for the superiority of his (*sic*) rational property.

The role of Amazon elites also marked the scenario of the development of the North region, based on the realisation and implementation of individualistic public policies and patronage. In this sense we can say that the causes of backwardness, impoverishment of the people, environmental degradation, social exclusion combined with the low quality of life in the Brazilian Amazon are the result of different factors and among them we highlight the predatory businessmen, bankers, agricultural interest groups, national and international traders and central government related to the patrimonial model of the colonels and landowners.

Currently, descriptions of the environment highlight its relevance and complexity, as the environment is seen in a global way interlacing the human being in the context of his or her life. In this way the individual is engaged in a collective process. Social movements have taken shape and technological changes are experienced in different contexts.

The paradigm change is preceded by a thorough formulation and institutionalisation of new problems. Everything indicates that this is the case for environmental problems: they gradually fill the political agenda, the philosophical and scientific fields, together with the technological and economic interests of global society. Due to complexity, the formulation of environmental problems in their systemic dimension seems to converge on sustainability: a problematic idea, a paradigm in construction, which is articulating the various dimensions of environmental issues, sectional interests, disciplinary efforts, social movements and the collective imagination. As if it were a heuristic guideline, the idea of sustainability summarises the many issues that plague our civilisation and guides the construction of perspectives and viewpoints for the integration of complexity in search of a new model or paradigm for global development.

Dealing with regional development implies a discussion about sustainability, its components and dimensions. Designing sustainability as a paradigmatic model for addressing social problems is a challenge. From this assumption the Amazon is not constituted in the correlation of the idea of sustainability in its complexity and originality, despite being the space on the planet that concentrates such complexity in a delimited geographic division.

The Legal Amazon

The creation of the Legal Amazon was guided by administrative, political and economic criteria aimed at planning. From its inception, this division endowed with natural and cultural complexity has been the object of an economic and administrative purpose, inspired by policies that focused on land occupation according to principles of a development model opposed to environmental issues.

In the late 1960s, the division of the Legal Amazon was incorporated by Law No. 5173 of October 27, 1966, by which the Superintendence for the Development of Amazonia (SUDAM) was created and to which was added a part of the state of Maranhão,

> 2nd Art. The Amazon, for the purpose of this law, covers the region encompassed by the states of Acre, Pará and Amazonas, the Federal Territories of Amapá, Roraima and Rondônia, and also the areas of Mato Grosso in the north of latitude 16, the State Goiás in the north of latitude 13 and the state of Maranhão at the west of the 44th meridian.

In 1977 the Legal Amazon was expanded by including the entire area of Mato Grosso, according to Article 45 of Complementary Law No 31 of October 11, 1977,

> Art 45. The Amazon region, referred to in Article 2 of Law No 5173 of October 27, 1966, also includes the entire area of Mato Grosso.

At the economic level some financial institutions stood out as, for example, the Bank of Amazonia (BASA) that is responsible for 82% of long-term investments and 52% of all loans applied by the public and private banking network in the North Region. It also provides loans which promote development, among them the unique FNO (special low rate loans for agricultural projects).

According to Monteiro and Coelho (2004), the political opening up in the 1980s increased political decentralisation and the formation of an organised civil society, which, through its historical resistance movement, triggered the claim for a set of demands that forced the discussion of sustainable regional development.

For many years, governmental actions have been implemented in Brazil through the Programmes for Local Integrated and Sustainable Development from the Federal Government and the Local Development Programme of PNUD (United Nations Programme for Development) encouraged by BNDES (National Bank for Development).

The Legal Amazon region delimits the outline of a paradox constituted by the existence of rich biodiversity, traditional cultures, local demographic pressures, extreme social and economic inequalities, emerging diseases, predatory occupation by agriculturists, conflicts over land, abundant water resources, the scene of controversial national policies, huge reserves of natural resources, and significant interventions of public works. In summary, there is a dynamic set of natural and anthropogenic factors that can equate the challenge of sustainability in its breadth involving biodiversity, cultural diversity and natural resources, all capable of being represented by the territorial issue.

In the environmental journey there is an emphasis on an empirical dimension that justifies the focus that the international community and various national segments plan for the Legal Amazon region: its undeniable environmental heritage, its biological diversity, inseparable from the inherent cultural diversity, its reserves of natural resources and its geopolitical location. However, this is part of a wider complicated reality embracing sustainability and socio-environmental tensions in a global context. The study of sustainability demands a return to the historical landmarks that have characterised the regulatory definition of the Amazon division.

In making the revision of the occupation policies of the Brazilian Amazon, Becker (2001: 135) notes that,

> Since the beginning of colonisation, exportation is the economic standard and the dominant motivation in the regional occupation. This dominance linked to the actual occupation of what is now the Amazon, Brazil and all Latin America, is an episode of the broad process of maritime expansion of European trading companies, formed as the oldest neighbourhoods of the capitalist world-economy. In other words, it is forged in the society-nature paradigm called "frontier economy" in which progress is seen as endless economic growth and prosperity, based on the exploitation of natural resources also perceived as infinite. The occupation of the Amazon was done in impulsive outbreaks, linked to the valorisation of products in the international market, followed by long periods of stagnation.

Most of the errors and mistakes made in the actions, programmes and social projects were due to a lack of knowledge of the historical experience. Neto (1979), a scholar of the Amazon, helps us review the various stages the Amazon went through and the development of its regional economy. In his words,

> Since the arrival of Vicente Yáñez Pinzón in 1500 until today, the Amazon has gone through several stages, and for each of them, the government actions were different.
>
> Thus, we can define: 1st) Period: 1500–1750 – Phase: Conquest (hinterland drugs) – Initial mark: Discovery of the Amazon by Pinzon. 2nd) Period: 1750–1850 – Phase: Occupation (farming business) – Initial mark: Treaty of Madrid. 3rd) Period: 1850–1946 – Phase One: Exploration (rubber cycle) – Initial mark: Creation of Amazonas Province. 4th) Period: 1946–1966 – Phase: recovery (planning) – Initial mark: Brazilian Constitution. 5th) Period: 1966 – Phase: Integration (reevaluation) – Initial mark: Operation Amazon (p.67).

Planning for regional development involves politico-parties' interests that do not always meet the interests of the region and much less satisfy the social and economic conditions of its population. According to Neto (1979: 26),

> The regional growth results from a range of decisions emanating from inside and outside the region and leading to inter-regional trade. Positive results depend on the region's ability to diversify its economic structure and minimise the polarisation effects exerted by the region-primate. The export sector will determine the infrastructure of the region during the early stages of regional development.

The factors of non-regional development indicated by research are: depletion of natural resources; structural change in demand; inefficiency of the politico-social structure; and economic and public administration dependence (when the municipal administration is subordinate to the state).

The State of Tocantins

Revisiting the creation of Tocantins in the context of regional development of the Legal Amazon is important in order to understand the UFT case. According to Cavalcante (1999), the movement towards the creation of Tocantins state has had separatist roots from the eighteenth century when it was still part of the Goiás context. Since 1820, the time of the Revolution of Porto in Portugal, the population of Goiás had been considering the ideals of freedom against the Portuguese occupation and exploitation developed by the mercantile logic. In the social construction of the population, college students were in a prominent position in the separatist struggle and in the 1960s the North Goiano Students' House (Cenog: *Casa do Estudante do Norte Goiano*) was created. The creation of the state of Tocantins was already among its goals. According to Santos (2007), one of Cenog's objectives was the search for social and economic improvements for the northern region of Goiás State which, in that context, was considered as abandoned.

> In a political context which emphasised development also encouraged by the government of Goiás state (Mauro Borges´ management), the college students of the state supported the separatist campaign in favour of the creation of the State of Tocantins. The achievement of political autonomy in northern Goiás was identified as a precondition for economic and social development of the region (Santos 2007: 41).

The state of Tocantins is located in the Northern Region, which occupies most of the Brazilian territory, with an area of 3,853,327.2 km^2 reaching 45% of the Brazilian territory, but in spite of having this vast territory, its population density is scarce, not reaching 8% of the population of the country. Its main feature is the Amazon rainforest region which is surrounded by the Amazon River basin with a hot and humid climate. Tocantins is one of seven states of the North and its population is made up of mixed race persons, descendants of Indians and slaves, and it receives a large migration of people from the South and Southeast of Brazil. The regional issue in

Brazil is a process that started with the flow of Portuguese migration from the colonial period in the early years of the eighteenth century with the discovery of gold in Minas Gerais. In the Amazon, for example, the territory previously occupied by the Indians is facing a messy economic exploitation that aims mainly at the accumulation of wealth. The part of Goiás, formerly located north of the state, is today the state of Tocantins. The territory of Guaporé corresponds to the current state of Rondonia. The area of White River corresponds to the current state of Roraima.

There was an expansion of higher education in the education sector: UFT (Universidade Federal do Tocantins), UNITINS (Universidade do Tocantins), ULBRA (Universidade Luterana do Brasil), FACTO (Faculdade Católica do Tocantins), FAPAL (Faculdade de Palmas), Universidade OBJETIVO (Objectif University), ITPAC (Instituto Tocantinense Presidente Antônio Carlos), UNIRG (Regional University Foundation of Gurupi), FIESPEN (Faculdade Integrada de Ensino Superior Porto Nacional), FUNDEG (Fundação de Educação Geni Nunes), FECIPAR (Faculdade de Educação, Ciências e Letras de Paraíso), FIESC (Faculdade Integrada de Ensino Superiorde Colin), Fasamar (Faculdade de São Marcos), UNEST (União Educacional de Ensino Superior do Médio Tocantins) were the 14 institutions in operation in the 1990s. Among those universities and colleges, one is federal and another one is from the state, with more than 100 undergraduate programmes and several post-graduate programmes. In this context, the increase in funding for the Northern Region is outstanding.

Events and committees promoted by the federal government have led to discussion of Amazonian issues at a national level. The Seminar of Post-Graduate Science and Technology in the Legal Amazon is an example of the attention that the government is giving to the region. There are also investments from the federal government in research institutions in the Northern Region to develop links with research groups in any region of the country associated with consolidated post-graduate programmes.

The increase in the number of scholarships for post-graduate programmes in the Amazon is a clear policy of the federal government to promote the region. Moreover, the Association of Amazonian Universities (UNAMAZ, Associação de Universidades Amazônicas), a multilateral agency for scientific and technological cooperation, non-governmental, non-profit making, and the Centre for Advanced Amazonian Studies (NAEA, Núcleo de Altos Estudos Amazônicos) of the Federal University of Pará (UFPA) are also noteworthy actors. Researches in development by UFT researchers at the Bananal Island, in the Ecological Park of Guangzhou are outstanding.

The Federal University of Tocantins

It is in this scenario that the UFT was established in 2003 with the commitment to contribute with dedication to the specificities of the Legal Amazon, which makes the University an instigator of sustainable regional development. The creation of the

UFT was contemporary with the very creation of the State of Tocantins and between 2003 and 2009 the UFT advanced steadily. Among the achievements, we can emphasise its institutional consolidation, the growing numbers of teachers and civil servants, courses and the number of places for undergraduate and post-graduate students in its seven campuses located in various cities in the state.

There is a recurrence of the UFT institutional commitment in its official documents regarding the sustainability of the Amazon. However, such an undertaking involves a series of conditions that suggest the complexity of environmental issues. The challenge for the new institution, which was created according to academic standards of the federal government, is to establish interdisciplinary activities aimed at promoting sustainability in the Amazon territory.

Regional Development

Regional development is a process of global social change, not only with an economic focus but also with a focus on cultural, political and social change. It requires the establishment of a new mentality of both the population and the rulers to reconsider the region where there are possible changes based on new directions. It consists not only in solving current problems but also in identifying new problems, considering that the role of any government is to raise the standard of the quality of life and to strengthen its political independence.

Some important indicators are taken into account in Neto's evaluation (1979) when referring to regional development, such as,

> a) that there is an increase in income per capita (including this condition is particularly important in the economy that has high rates of population growth), b) that there is an improvement in income distribution, or at least it is not worsening, c) that the absolute number of people below a certain level of real income (the subsistence, for example) is decreasing, d) that regional differences of income per capita in relation to national average are not increasing (p.29).

Sustainable regional development is a challenge for the regional population and for any local government seeking to outline and achieve three major goals that we consider fundamental in this process: economic growth; equity (fair distribution in the economic and social fields); and protection of the environment. In theory, regional development must ensure the quality of life. Despite the HDI (Human Development Index) of the UN, the quality of life is measured by the real possibility of people being able to meet the basic or fundamental needs of their integral development that goes far beyond what is historically accepted in the poorest regions, namely those related to subsistence.

In contrast to the prevailing model of capitalist development, the regional development in which we believe is constituted in the project of social movements which from their demands define a model that: emphasises collective participation and is geared to meeting the basic needs of the population of the region; is self-independent, endogenous, since part of the interests and the real possibilities of each region are inspired by cultural values; is self-sufficient as it counts on its own resources and with the richness of its forces at each step; ecologically speaking, is good and right, preserving the environment and the natural resources; and based on changes in attitudes and social transformation.

In order for the region to face the challenges of regional development, higher education has an important contribution to make, dependent on its scientific capacity. But with regard to researchers in science and higher education in the Amazon, in particular, Aragón (2001) argues that developing countries must confront the critical factors that hinder the construction of their own scientific capacity,

> Challenge 1) Reduce inequities within developing countries themselves; Challenge 2) Redirect the scientific research to the interests of developing countries themselves; Challenge 3) Invest in education and scientific research without undermining the social policies that address the basic needs of the population; Challenge 4) Monitor the brain drain; Challenge 5) Refocusing international cooperation; Challenge 6) Incorporate the popular knowledge in scientific research; Challenge 7) Improving interdisciplinary research; Challenge 8) Strengthen networks; Challenge 9) Responding to globalisation; Challenge 10) Formulate and implement legislation relating to research and intellectual property rights; Challenge 11) Improve management; Challenge 12) Strengthen new leaders (p.4).

For any development plan for the Amazon and thus for the North Region, there is a need to determine the extent to which teaching, research and extension contribute to improving not only economic development but also the quality of life in the region. It is also important to measure how higher education has produced consistent knowledge and not a mere transmission and uncritical transposition without proper contextualisation of the local reality.

In the discussion on public policies for sustainable development of Tocantins, which has been the discourse of the state government, the government proposition revolves around ecological-economic zoning as a tool for planning and regulating the use of the land, aiming at territorial transformation based on the recognition of spatial and temporal differences. According to Zitzke (2004: 86),

> …the official discourse assumes a new development strategy based on the articulation of economic, social and environmental dimensions, through land management that involves shared decisions of government and civil society on the sustainable use of natural resources in a given area. The analysis of the proposed sustainability shows that the relationship never really happened, because the public events always prioritised international private capital, camouflaged as a promoter of sustainable development. This capital is so exploitative, or more, because it does not consider the specificities of places and for having nowhere else to expand, joins the state to take ownership of their natural resources through the implementation of major projects of development.

Another issue that should enter the debate is that not all growth is constituted under development. In an analysis of the government proposal for sustainable development in the State of Tocantins, Zitzke (2004: 94) wrote,

> The economic development initiatives, which more properly should be called economic growth, in Tocantins, demonstrate the opposite of the discourse, because in many cases the ecological and social dimensions are ignored in favour of the economic dimension. This fact can be observed from the social and environmental impacts resulting from major economic development projects such as Luis Eduardo Magalhães Hydroelectric Plant and the

relocation of families from the coastal processes of compensation based on the valuation of material goods and natural "goods" from an economic view of valuation of nature.

Concerning the development process, this is a one-dimensional concept that presents the economic factor as the crucial aspect of development guided solely from material and technological angles. The parameters for quantifying the development in this conception are based on GDP – Gross Domestic Product – and Per Capita Income. This view gives absolute importance to the material dimension, reducing the process to the financial aspect at the expense of the complex citizens' needs chain.

When we speak of the regional term, we take as the basis a specific region where people share a territory, a history, a culture and a common destiny. The term sustainable means that it benefits future generations, ensuring them an economic and social development through responsible conservation of nature. Discussing what is being called sustainable development or eco-development involves many currents, around which is built an ideological discourse apparently critical, but mostly does not formulate an alternative policy to the actual model of society based on capitalism.

Environmental degradation and its relation to the reduction of the standard indicators of quality of life are identified amidst this panorama. This is especially marked in urban areas since the late 1990s with the aggravation of problems due to the lack of basic sanitation which is, for example, an indispensable condition to stop epidemics spreading. Besides this, there is an increasingly accelerated formation of pockets of poverty, depending on rural exodus, that are often located in suburbs and without basic infrastructure, being large shanty towns in areas with life-threatening risks.

What we are considering in this discussion on sustainable development is the possibility of reconciling economic development with the enhancement of life on the planet, which requires income distribution and discussions about environmental problems grounded in the social and cultural values of the ethical field.

The discussion on sustainable development has strengthened since the organisation by the UN in 1972 of the Stockholm Conference on the Environment through to Eco–92 (World Conference on Environment and Development held in Rio de Janeiro, Brazil in 1992). From this movement two myths were conveyed: first, that the poor are the main cause of the destruction of the so-called environment and, secondly, that population growth in the southern hemisphere is the main determinant of environmental degradation. Social development and the new world economic order are fronts that need to be discussed together, and therefore follows this discussion of regional development.

By locating the UFT in the socio-environmental scenario of the Amazon, we cannot fail to refer to the unity between biodiversity and cultural diversity, highlighting the paradigmatic issue of the organisation of the protected areas of Amazonia. According to 2009 data from the ISA (Socio-Environmental Institute), the Brazilian Amazon has 309 federal and state conservation units and 405 indigenous lands, the latter occupying 21.69% of the Amazonian territory.

Indian rights were recognised in 1988 by the Constitution of the Federative Republic of Brazil, as follows,

> Art. 231. Indians shall have their social organisation, customs, languages, beliefs and traditions, and rights on the lands they traditionally occupy, incumbent upon the Union to demarcate them, protect and enforce all of its assets.
> § 1 – Lands traditionally occupied by Indians are those inhabited by them on a permanent basis, those used for their productive activities, those indispensable to the preservation of environmental resources necessary for their well-being and for their physical and cultural reproduction, according to their uses, customs and traditions.
> § 2 – The lands traditionally occupied by Indians are intended for their permanent possession, leaving them the exclusive enjoyment of the riches of the soil, rivers and lakes existing therein.
> § 3 – The use of water resources, including energy potential, exploration and mining of mineral wealth on Indian lands can be effected only with the authorisation of Congress, after hearing the affected communities. This right is guaranteed by law.
> § 4 – The lands in this article are inalienable and unavailable, and the rights not prescribed.
> § 5 – It is forbidden to remove Indian groups from their lands, except by Congress referendum, in case of disaster or epidemic that endangers the population, and in the interest of the sovereignty of the country, after approval by the National Congress, ensuring, in any case, the immediate return as soon as the risk ceases (Constitution of the Federal Republic of Brazil. 2010: 46).

After a long period of apparent stagnation, the indigenous population is growing. According to the IBGE (Brazilian Institute of Geography and Statistics), the percentage of Indians in the Brazilian population (147 million) was 0.2%, totalling 294,000 Indian people, in 1996. Also according to the IBGE, in 2000 the percentage rose to 0.4% of the Brazilian population (183.5 million), totalling 734,000 Indian people and according to current projections of the FUNAI (Indian National Foundation) the Brazilian indigenous population should reach one million in 2010, an amount comparable to the entire population of the state of Tocantins in 2009.

The indigenous issue cannot be separated from the environmental dimension. Indigenous lands form a space defined not only by local cultures but also by the conservation of natural resources in the Amazon. The Federal Constitution guarantees to the Indians the enjoyment of natural resources in their territories. However, there are two aspects of the impacts related to indigenous people: the impact of modernisation and the cultural invasion of indigenous lands for the exploitation of natural resources. The change in the form of occupation of the Amazon territory since 1970 with the construction of highways and infrastructure works has produced intense pressure on the territories. The contact and assimilation of central culture unmediated by the means of communication and social relations have introduced a constant breakdown of the Amazon's local people cultures, in particular those of indigenous people.

Regardless of discussions about the legitimacy of the existence of indigenous lands, we are facing as a matter of fact in the rule of law the possession and enjoyment of indigenous lands by indigenous people, as part of the constitutional provisions of 1988, though little is known of this to most Brazilian citizens. We have a view that civil rights must be protected and preserved as are many others contained in the Constitution. Such considerations are integrated into the scenario of the creation of the UFT, outlined in the Legal Amazon and circumscribed by the creation of the state of Tocantins. According to the IBGE Census 2000, we have a self-declared indigenous population of 10,581 persons in Tocantins, that is, 3.6% of the indigenous population in Brazil.

We give some prominence to indigenous issues because of the contributions they can make to the central theme. First, the inseparable unity between local cultures and environmental diversity represents a case study of discussions of socio-environmental currents that are typical all over the world, the relationships between local culture and environmental sustainability. Secondly, we have seen the progression of conflicts among the interested parties on indigenous lands, conflicts showing an exponential escalation with serious consequences for the political and environmental balance of the region. Thirdly, there is the creation of the UFT as a space responsible for the production of knowledge, the intentional mediation of culture and the training of citizens committed to the sustainable development of Amazonia.

Several actions for indigenous issues have been institutionalised by the UFT among which can be highlighted the creation of the Centre for Studies and Indigenous Affairs and programmes to support indigenous students. One of the great challenges of indigenous education is the effective recognition of cultural diversity or ethnic origin, including the construction of processes that allow mediation between local cultures and our core culture – academic, technical and scientific. However, the UFT, because of its location, organisation and institutional motivation, is placed in the epicentre of those regional issues related to cultural diversity, biodiversity and environmental conflicts.

Using the rule of law to build our scenario, we have to mention another type of protected area which consists of the Conservation Units. Once again we are faced with structures quite unknown to most citizens, including those who constitute our intellectual elite. Demonstrations against the maintaining of 'protected green areas' are recurrent at the expense of economic and material advancement of society, as if both were mutually exclusive and opposing elements. Let us see then, in a synthetic manner, what the Conservation Units are and how they fit into our scenario.

Conservation Units are referred to in Article 225, dedicated to the environment, of the 1988 Constitution. Law No. 9985 of July 18, 2000 brings regulations to Art. 225, § 1, paragraphs I, II, III and VII of the Federal Constitution. This law established the National System of Conservation Units of Nature (SNUG) and established the criteria and standards for the creation, implementation and management of the Conservation Units,

Article 2nd. For the purposes of this Law, it is meant by:

> I – conservation unit: territorial space with its environmental resources, including territorial waters with relevant natural characteristics, legally established by the Government, with conservation objectives and defined limits, under special administration, to which are applied appropriate protection guarantees;
> XI – sustainable use: exploitation of the environment in order to perennially ensure the renewable environment resources and ecological processes, maintaining biodiversity and other ecological attributes in socially fair conditions and economically viable;
> Art. 3rd The National System of Nature Conservation Units – SNUG is made up of all federal conservation units, state and municipal governments, in accordance with the provisions of this law.

Thus, the principles of the Constitution regulated by Law No. 9.985/00 promote the establishment by the government of the territories described for environmental conservation, delimiting biodiversity, water and natural resources in general. The legal resolution in Article 7 also categorises the protected areas into two types,

> Art. 7th The conservation units, member of the SNUG, are divided into two groups with specific characteristics:
> I – Integral Protection Units;
> II – Units of Sustainable Use.
> § 1st The basic goal of Integral Protection Units is to preserve nature, admitting only the indirect use of its natural resources, except in cases provided by this Law.
> § 2 The basic objective of the Sustainable Use Units is to combine nature conservation with sustainable use of a portion of its natural resources.

Conservation Units represent another type of qualified division, involving approximately 40% of the Legal Amazon region, which suffers the enormous impacts of a development driven by the predatory model that presupposes an incompatibility between economic development and the sustainable use of natural resources. However, the establishment of Conservation Units is an important element in the designation of the rule of law enshrined in the Constitution, integrating the structure of rights and duties inherent in the full use of citizenship.

In brief, the creation of the UFT can be located in two areas qualified by environmental law according to the constitutional apparatus: the indigenous lands and conservation units, both of which are under pressure from a development model in crisis that brings serious consequences and conflicts in the Amazon region. Incidentally, we have introduced two components that are inseparable from the point of view of sustainability: the dialectic between cultural diversity and biodiversity.

The UFT and the Challenges of Education for Sustainable Development in Amazonia

One of the most intriguing aspects of human thought consists in building models to understand reality, integrating complexity under temporal and spatial conditions. Global society is currently facing the challenge of building paradigms that promote the reintegration of knowledge that has been divided and segmented over the past 400 years. This challenge is central to the construction of paradigms of sustainability. It is no exaggeration to say that the Legal Amazon is an exemplary case at a planetary level.

The modern ideal of social and economic production and reproduction based on the dominance of the forces of nature is in question. Since the advent of the modern scientific revolution, scientific and technical society has been organised around an explicit opposition to nature that involves knowing how to master it. Modernity has brought efforts of demarcation between religion, philosophy, scientific knowledge, technology and politics. Progressively, from the sixteenth century we have witnessed in the Western world the formation of disciplinary fields; the segmentation of the natural sciences culminated in the separation between the human and social sciences in relation to natural science in the late nineteenth century. The interest in knowing to understand the sense of reality and giving a meaning to the exercise of power became an interest in controlling natural forces and later the human and social forces.

The ways in which people establish their relations with nature are inseparable from the ways they interact and produce their culture. Our model of development went into crisis and with it went the very meaning of our established knowledge. According to Martins and Melo (2007: 93),

> Putting it in simple words, sustainability is providing the best for the people and the environment both now and in the near future, undefined. The approach we obtain from sustainability leads us to the action of setting up the civilisation and human activities, in a way that society, its members and their economies can fulfil their needs and express their greatest potential in the present while preserving biodiversity and natural ecosystems, planning and acting to achieve proficiency in maintaining these ideals indefinitely. The human as a subject promotes his choices in the perspective of thought construction, considering the contact with new values, cultures and knowledge based on ethical principles.

This quotation refers to the configuration of new cultural paradigms in line with formal education. Education is a cornerstone of Western civilisation, a condition for the production of technical and scientific rationality, the dialectical process of self-construction of our culture, the means by which society can share the solutions of problems that jeopardise its own existence in the near future. Education is a human process, complex, constitutive of its own social system and the culture of our civilisation. Greek thought was already conceptualising this unity through the concept of *Paidéia* as a process of complete training and self-training which aims at the full exercise of citizenship. That is, the imposition of the Greek polis is inconceivable without the *Paidéia* and its ideals of education, democracy and political coexistence.

Only this type of education may apply properly to the word *formation*, as Plato used it the first time in a metaphorical sense, applying it to educative action. The German word *Bildung* (formation, configuration) is the one that designates in a more intuitive way the essence of education in the Greek and Platonic sense. It contains at the same time the artistic and plastic configuration, and image, 'idea' or 'type' rules to be discovered in the intimacy of the artist. In every place where this idea reappears later in history it is a legacy of the Greeks and appears when the human spirit abandons the idea of grooming on the basis of exterior purposes and reflects on the essence of education itself (Jaeger 2003: 13–14).

We are looking for a new *Paidéia*, no longer guided by the principles of an absolute metaphysics but by the dynamic construction of an intersection between the fields of human reality that have been separated by technical, scientific and material advancement. The idea of 'environmental education' was an archetype of this effort of reconciliation in pursuit of a full training.

The role of higher education in sustainable regional development is important because the performance of a university through the triad 'teaching-research-extension' is constituted as:

a. one of the main pillars to support economic, social and cultural development;
b. an important promoter of development and progress;
c. the privileged locus where the academic community interacts, develops and acquires knowledge and skills in the quest to understand and intervene in the reality in which it is located;
d. where the solution to problems and challenges of the economic-social context is investigated; and
e. for the training of qualified personnel, etc.

From this assumption, when defining the role that a university can take in relation to regional development, there should be considered the diversity that affect: (a) the region (its characteristics, needs, potential areas, culture, among other factors); (b) the university (its characteristics, its regional vocation, capacity of its installations, material resources, staff, its investment); (c) government policy; and d. the institutional policy. In this projection the university can contribute to the region through:

1. supporting the creation and development of small and micro-businesses, cooperatives management;
2. technical support in preparing and implementing action, regional programmes and projects;
3. formulation of policies for the insertion of marginalised and excluded minorities;
4. technology transfer;
5. business and trade management, university-enterprise integration;
6. continued training and staff specialisation in the region;
7. technical assistance and advice;
8. establishing technical cooperation agreements;
9. providing information services, publications and communications;
10. organisation of exhibitions, scientific congresses to disseminate research;
11. promotion of artistic and cultural activities and entertainment;
12. teachers and students interchange with other IES (Higher Education Institutions) in the region and abroad;
13. support to public interest campaigns;
14. training of agents to work in specific socio-economic projects.

The university is not only an architectural project and it is not present in society only because it brings economic development to a city or region. Beyond the utilitarian view, the largest contribution of higher education is to train people skilled enough to create new knowledge and find alternative solutions that prioritise human well-being. Thinking in regional development is, before anything else, thinking about the participation of a local society in the ongoing planning of space occupancy and in the distribution of the fruits of the growth process.

The development of a region can be explained as a result of the interaction of three forces: the allocation of resources (regional participation in the use of national resources and state); economic policy (the action of the central government that can affect the region positively or negatively); and social activation (creation of a number of political, institutional and social factors). Sustainable regional development depends on the reconciliation of policies that boost growth with local objectives. The organisation of local society can transform the positive effects in growth, or better, in development for the region. And the region cannot be seen merely as a geographical factor, but as a social actor, as a living element in the planning process. The state is the one which sets the rules of the game while the region is the negotiating part which must insert itself into decision-making mechanisms to make agreements, transactions, resolve conflicts, and finally it must have the ability to transform the external impulse of economic development into development with social inclusion. The solution of regional problems and therefore the improvement of the quality of life require the strengthening of society and local institutions as they are the ones that will transform the external impulse of growth into development.

Conclusion

From this panoramic vision of the main lines of a systemic view, we stress that the state of Tocantins retains, since 1993, the Commission for Environmental Education of the State of Tocantins, established by Decree No. 8629 of August 16, 1993. Law No.1374 of April 8, 2003 established the State Policy of Environmental Education, followed by deployment of the State Programme of Environmental Education that regulates it. The State Policy of Environmental Education of Tocantins was created in line with national policies established in the following legal provisions: Article 225 of the Federal Constitution, Law 9795 of April 27, 1999 establishing the National Policy for Environmental Education and Decree No. 4281 of June 25, 2002, regulating Law No. 9795 of April 27, 1999.

In conclusion, we propose a set of challenges related to sustainability education from the creation of the UFT within the Amazon:

a. to contribute to the construction of a new paradigm that allows a reasonable understanding of the complexity of the Amazonian division, integrating anthropogenic and natural dimensions, according to an interdisciplinary point of view and the preservation of cultural and regional biodiversity;
b. to promote the full formation of citizens who can act for the benefit of the sustained and sustainable development of the Amazon, citizens who would be able to familiarise themselves with the multiple diversities and be aware of the impacts of unconditioned modernisation;
c. to produce and disseminate knowledge to promote the economic and social development of the region in all strategic areas of the territorial division of the region, without losing sight of the adoption of sustainable solutions, renewable resources, clean technologies and balanced relations with the environment;
d. to recognise the protected territories and promote actions in favour of sustainability in its broadest sense: the Conservation Units and indigenous lands as

permanent elements of the institutional, scientific, cultural and technological scenario of UFT;

e. to achieve social and environmental mediation in the region through education, research and extension;
f. to integrate institutional actions with the National and State Environmental Education policies; and
g. to construct, propose and share solutions for sustainability of urban and rural areas, that are under pressure from the following factors: impacts of unsustainable activities, disorganised occupation of the territory and socio-economic inequalities.

According to the 2009 IBGE population projections, 116 municipalities of Tocantins state have less than 10,000 inhabitants, nine cities have between 13,000 and 15,000 inhabitants and only 14 have between 14,000 and 189,000 inhabitants. Therefore, the creation of the UFT has produced an area with a number of people much higher than most counties in the state of Tocantins. The majority of these inhabitants of the UFT area will have achieved educational qualifications which will be upgraded, multiplied and spread across generations. As we noted earlier, the UFT is a space created for one purpose: to generate knowledge, promote solutions and educate citizens in an integrated process aiming at sustainable development of the Legal Amazon Region.

The challenges mentioned above have been formulated only in 2010 because the UFT took the initiative to establish commitments to the sustainability of Amazon through its main regulatory frameworks, actions and institutional priorities. It is from this qualified area that we locate and formulate the challenges to be faced in the coming years for the next generations who are being prepared at the UFT.

It is in this context of advances and reverses that the important role of higher education is relevant in facing the issue of regional development when considering that the university creates the best strategic conditions to outline and forward the discussion of social and environmental problems, especially, in the region in which it operates. Thus, we believe that sustainable development is a comprehensive process that involves an integral way of reading the reality and acting ethically on it in a way that it cannot be limited only to subjects as it is included in everyone's life.

That is how we understand higher education, with the UFT case, as institutions that induce sustainable regional development issues. The university has the potential to create the strategic conditions to outline and forward the discussion of social and environmental, political and educational problems, especially of the Legal Amazon, in which it operates.

References

Aragón, L. E. 2001. *Ciência e Educação Superior na Amazônia: Challenges and opportunities of international cooperation*. Belem: UNAMAZ (Association of Amazonian Universities) NAEA (Nucleus for Amazon High Studies).

Becker, B. K. 2001.Models and scenarios for the Amazon: the role of science. Reviewing of the Amazon occupation policies: is it possible to identify models to project scenarios? *Revista Parcerias Estratégicas*, no. 12 (September).

Brazil. 2010. *Constitution of the federal republic of Brazil.* p.46. Brasília: Publication Office.
Carril, C. 2007. *Cultura tecnológica sustentável: A case study of the cognitus in petrobras project.* São Paulo: Anhembi Morumbi Editor.
Cavalcante, M E S R. 1999. *Tocantins: Separatist movement in northern Goiás (1821–1988).* São Paulo: Anita Garibaldi.
Jaeger, W W. 2003. *Paidéia: The formation of the Greek man*, 4th ed. São Paulo: Martins Fontes.
Melo, C. K. and Martins, J. R. 2007. Dimensions and sustainability. *Revista Amazônia Legal de estudos sócio-jurídico-Ambientais,* January-June, Year 2, no. 3: 93–103.
Monteiro, M. A. and Coelho, M. C. N 2004. Federal policies and spatial reconfigurations in the Amazon. *Pará: Novos Cadernos NAEA*, 7(1), Available at http://www.periodicos.ufpa.br/index.php/ncn/article/viewFile/38/32 (accessed October 10, 2009).
Neto, M J. 1979. *O Dilema da Amazônia.* Petrópolis: Vozes.
Santos, J S. 2007. *O Sonho de Uma Geração.* Goiânia: ECG.
Schramm, F R. 1999. *A moralidade das biotecnologias. I Brazilian congress on biosafety.* Rio de Janeiro: ANBio.
Souza, M. C. S. 2008. *Bens públicos e externalidades.* http://introducaoaeconomia.files.wordpress.com/2010/03/texto-bens-publicos-e-externalidades.pdf (accessed January 20, 2009).
University of Tocantins. 2003. *Institutional development plan.* Palmas: PDI–UFT.
Vieira, L (ed.). 1997. *Cidadania e Globalização.* Rio de Janeiro: Record.
Zitzke, V. A. 2004. O discurso do desenvolvimento sustentável no Estado do Tocantins. *Interface. NEMAD (Centre for Education, Environment and Development) UFT,* 1(1), Porto Nacional.

Part III
Trends and Challenges of Educational Provision for Sustainable Development

Chapter 11
Social and Environmental Design as an Educational Pedagogic Resource for Sustainable Development: An Experience with NGOs and Universities

Nara Silvia Marcondes Martins, Ivo Eduardo Roman Pons, and Petra Sanchez Sanchez

Introduction

In this chapter we report the work of a university extension project *Design Possível* supported by the NGO *Design Possível*. Since its inception in 2004 this academic project has been concerned with experimental practice in the socio-environmental design production and sustainability field. It has also focused on research into the development of designers as society transformation agents in the perspective of the new socio-economic order, which means training designers for the twenty-first century.

In the project, design with other practical applied working fields has a number of elements, including research synthesis, critical reflection, creativity and sensitivity that are always present in a designer's professional activity. Indeed, the project embodies much of the designers' actions and therefore it is emblematic of the profession, as are its study and analysis.

The International Council of Societies of Industrial Design (ICSID) (2010) describes design as:

> A creative activity whose aim is to establish the multi-faceted qualities of objects, processes, services and their systems in whole life cycles. Therefore, design is the central factor in the innovative humanisation of technologies and the crucial factor of cultural and economic exchange www.icsid.Org/about/about/articles31.(Accessed: March 9, 2010).

In the environmental damage context, this description reminds us of Manzini's argument (2002) that the depletion of natural resources requires a new mindset for the twenty-first century. Goods manufacturing needs methods and materials that

N.S.M. Martins (✉) • I.E.R. Pons • P.S. Sanchez
Mackenzie University, São Paulo, Brazil
e-mail: petrasanchez@mackenzie.br

M.L. de Amorim Soares and L. Petarnella (eds.), *Schooling for Sustainable Development in South America*, Schooling for Sustainable Development 2,
DOI 10.1007/978-94-007-1754-1_11, © Springer Science+Business Media B.V. 2011

cause less environmental degradation. The role of design is to offer alternative solutions to innovate with in the material culture and thereby improve the local culture and align the behavioural culture. It has been demonstrated that design can be responsible for other connections between user and object without compromising its functionality (Pons 2006).

The project *Design Possível* originated from the Projeto de Cooperação Ecodesign Social Brasil-Itália (Social Ecodesign Cooperation Project between Brazil and Italy) that originated in the development of academic extension activities involving international collaboration. The proposal was to produce objects in partnerships with private companies and the third sector, seeking to promote social and community development. The project involved students, researchers and professors of industrial design at Universidade Presbiteriana Mackenzie, São Paulo (Brazil) and Universita degli Studi di Firenze, (Italy) and three non-governmental organisations (NGOs) in São Paulo working with industrial and household waste.

With international cooperation, the project was primarily aimed at creating design projects related to household objects. The project methodology had as prerequisites the conversion of industrial waste into raw materials and the use of technology and production processes that were commonplace for artisans. This was an appropriate way to promote social inclusion of people from poor communities who worked with the reuse of recyclable waste (wood, textiles, canvas, among others). It also sought to promote a relationship between project participants and design students from both countries with the purpose of contributing to methodological concepts in order to facilitate trade in these products in Brazilian and Italian markets.

As work developed, the group became stable and began to design different types of production processes involving excluded communities and companies. The purpose was to design ideas and concepts and create objects in line with quality of life issues in terms of ecology that were socially just and economically viable. The principles of operation of *Design Possível* are relevant for sustainable development proposals since they encourage environmental education and the development of socio-environmental design.

Philippi Jr and Pelicioni believe that environmental education is a process of political education that enables the acquisition of knowledge and skills, as well as attitude development that will necessarily become a citizenship practice to ensure the maintenance of a sustainable society (Philippi and Pelicioni 2000). Sorrentino et al. (2005) assert that, "environmental education, specifically, when encouraging citizenship, can create the possibility of a political action, helping to generate a collective that is responsible for the world it inhabits".

Private companies interested in associating their brands and promotion projects with socio-environmental responsibility concepts were very important to consolidate the *Design Possível* Group. They were also important for the technical and professional development of individuals' skills in the NGOs which created the products and promotion material. In this sense, they also tried to generate a competitive edge with their clients and customers. In Brazil, some companies have realised the importance of investing in environmental protection and training for their employees in compliance with environmental laws and guidelines of the

Environmental Education National Policy. Law No. 9795/99, Article 13, deals with non-formal environmental education and highlights the importance of developing actions and practices aimed at raising awareness and the educational organisation of the community on environmental issues together with participation in protecting environmental quality (Brazil 1999).

It is worth pointing out that the partnership between *Design Possível* group's designers, private companies and NGOs allowed the achievement of projects which use recyclable materials in the production of ecologically correct promotional material that incorporates three important concepts; design, ecology and social responsibility. This partnership promoted exchanges among the three groups involved and these have furthered the pedagogical actions of researchers, professors and university students in poor communities. Actually, the communities were sensitive to the importance of using recyclable materials. Individuals saw the need to become professionals or to learn how to make products and craftwork which would contribute to the recognition of their work and to the maintenance of their organisations. They were equally sensitive to the fact that they can benefit the environment. In addition, it should be emphasised that, besides the knowledge passed on regarding design, ecology and recycling projects to the participants in the communities involved, academics and researchers were concerned to foster greater interaction between NGOs and the university and aspects related to the education of citizens.

Starting Point

Design Possível began in 2004 in São Paulo in the Industrial Design programme of the Universidade Presbiteriana Mackenzie (Mackenzie Presbyterian University) and Università di Firenze (University of Florence) in Italy. It was initiated in uncertain times, due to economic and financial globalisation and awareness of global environmental problems. The goal was to develop social and ecological actions between two different societies, Brazil and Italy. These actions focused on the development of projects for household objects with international cooperation using, as a prerequisite, industrial waste as raw material.

The objects designed were unique and made with solid waste using two different manufacturing processes. It was an effective idea to develop research between two universities in different continents with contrasting realities. They shared the same concerns about preserving the environment and were interested in social environmental responsibility issues. With the foundations of Projeto de Cooperação Internacional Ecodesign Social Brasil-Itália laid, it was agreed that the designer's role today is to work as a mediator between the market and the producing community. This bond was then named *Design Possível* in Brazil and *Design Possibile* in Italy.

The participants in *Design Possível* sought to encourage students – future designers – to put into practice the words of Papaneck (1995: 52), "Design education in the twenty-first century should be based on ecological methods and ideals". Such education highlights, among others, the protection of biodiversity, the study of environmental sustainability, the environmental certification and reuse of waste. The solid waste reused not only can help reduce environmental impact but also be decisive for social inclusion and income generation in marginalised communities that inhabit the poorest regions.

Initially, the project development involved three non-governmental organisations in São Paulo: the NGO Florescer located in the slum Paraisópolis, the second largest slum in the city and the fourth largest in Latin America; the NGO Aldeia do Futuro, located in Americanópolis, south of the city; and the NGO Monte Azul that originated in Monte Azul, Peínha and Horizonte Azul, the slums of the south. All used as raw material waste that was either donated or purchased. It is appropriate here to provide some information for each of these NGOs.

The NGO Florescer – This was established in 1990 in São Manuel, São Paulo. In 1995 it was transferred to the city of São Paulo based in the slum Paraisópolis. The institution runs dance, music, soccer, English, fine arts and drama lessons that aim to educate children, teenagers and young adults. The NGO serves about 700 children in the different courses, promoting access to new knowledge. Among the various initiatives undertaken in partnership with *Design Possível*, the Recicla Jeans project stands out. It began in NGO Florescer, in partnership with the City of São Paulo and UNESCO. Its goal was to teach the art of sewing and modeling, offering the opportunity for professional development through this learning. Currently, youngsters and mothers from the community produce clothes and accessories with jeans patches that are donated by industries or, in the case of second-hand pieces, individuals. These products are marketed in Shopping Centre D in São Paulo by the NGO itself.

The NGO Aldeia do Futuro – This was founded in 1993 by entrepreneurs committed to the social problems related to education, culture and citizenship of young people and poor communities. It currently serves more than 500 young people who attend the courses the institution offers free of charge since their goal is to educate and train aware and involved citizens, as well as self-sufficient professionals. In 1998, the NGO implemented the project "Aldeia de Mulheres," where women have been offered professional training in handicrafts, which has given them the opportunity to produce bags, pillows, rugs, among others, all with recyclable material, thus generating extra income for the families involved.

The Associação Comunidade Monte Azul – This was officially founded in 1979 as an initiative of a pedagogue (Ute Craemer) who started having children from the slums in her own residence. Paper recycling workshops, the reuse of furniture and household objects, production of dolls, carpentry and computing were implemented. From the craft production of paper and wood income was generated which helped to maintain the NGO, including nurseries and day care centres, in addition to cultural activities in dance, music and drama. It serves approximately 1,251 children and adolescents in education and assists 4,000 people with health care.

Ten Brazilian students of industrial design studying at Universidade Presbiteriana Mackenzie were selected to visit these communities. They conducted a data and images survey, recorded the NGOs' activities so that Italian students and teachers could understand the life of the poor communities in which the NGOs were located.

All the data were examined and then edited, catalogued and systematised. In total, 14 CDs were sent to Italy, in addition to samples of processed and Tighten under-processed waste materials. After the research phase came the development proposals for manufactured products. Seeking to achieve effective cooperation among students, working pairs were created with one student from each university.

During the product development process, there were two weekly meetings on pre-arranged days and times. Initially, scraps and thumbnails were requested, which were then presented and discussed among everyone, always supervised by the coordinating professor. The ideas were exchanged via e-mail. The projects should represent at least a suggestion from each participant, a minimum of two products per pair, always between a Brazilian and an Italian. Some participants had made more than two products. The different types of the material shown – like jeans or wood – using craft techniques and the number of proposed themes caused different reactions in the Italian and Brazilian students.

The prototypes developed in the previous step were designed by the earlier mentioned NGOs for approximately 45 days. The ideas and concepts of the Italian participants' projects appeared difficult to implement. However they were more interesting than the Brazilians'. These projects contained computer and analog illustrations, sometimes photos of models or miniature prototypes. However, they showed no technical, production or construction details. A major concern for the Italians was the elaboration of concepts whereas, for Brazilians, it was the construction of the units that was most important, a fact that in some cases caused conflict between the production viability and the maintenance of the project ideology. However, the small amount of evidence of conflict disappeared with the realisation of the products as students became aware they had not only to solve problems but also to 'create solutions'. So, after each new concretised project the aesthetic and creative differentials were recognised. The outcome of this strategy pleased everyone. In Milan the work was exhibited in April 2005 at Fuori Salone, at Brasilartes Gallery and at the Brazil-Italy Institute.

A project of this type which contains a relationship between research and practice could be implemented in other poor regions of Brazil. In fact the student was engaged with the third sector, involved in the production of objects and rethought and assumed his or her role as a citizen-leader-educator and fulfilled the role of a designer, author and well-being articulator. He or she was also directly connected to social actions. The students, as future professionals, could then realise the importance of their work for society and their transformation potential.

The results of this initial activity of *Design Possível* demonstrated that the project comprised a group that promoted socio-environmental design. It took first place in the Planeta Casa competition in 2005; took part in the 1st Bienal de Design in São Paulo, and was the winner of Planeta Casa in 2006 in the Student and Social Action category. Several exhibitions were arranged in Brazil, such as Mostras de Boas Práticas Ambientais of the Secretaria Municipal do Verde e do Meio Ambiente (Exhibition of Good Environmental Practices of the Municipal Secretariat of Green and Environment) of São Paulo, which took place in 2006 and 2007 in the Ibirapuera Park in São Paulo. In November 2008 the Exhibition at Museu da Casa Brasileira was held under the title "Eu Não Sou Mais de Plástico, Sou Sustentável e Gero Renda", (I'm no longer plastic, I'm sustainable and generate income) with a combination of more than ten different groups involving professionals, students and NGOs. Abroad, *Design Possível* took part in Salão do Móvel in Milan (2006 and 2007) alongside the Brasilartes gallery and IBRIT Italy-Brazil Institute. It also took part in the 2006 Bienal Internacional de Design in Saint-Etienne, France.

Over the years *Design Possível* has united people with different intellects and knowledge – the traditional and the technical. It has mobilised people in order to produce objects which were important elements of social changes and to generate income for their producing communities. In this sense, its relations go beyond product development; they remain in the actions and experiences of all of the individuals involved. It is a process that gradually transforms the designer, craftsman, university, student and the know-how of the third sector.

Design Possível Project Consolidation in an NGO

In mid-2007 other objectives were planned within the group of *Design Possível* members. In fact, based on the close accumulated working experience with NGOs and productive groups and the growth and training of those involved, a need was identified to build a model that would allow the project to go well beyond the walls of the university. A project was started to convert the group into a formal association and this materialised in January 2009 with the founding of the Associação Design Possível.

With almost five years of experience, *Design Possível* has promoted interaction between design students and NGOs in developing products with production management. The group's work was founded on a partnership between the school and the community as it had provided for its self-sustainability, had enabled the exchange of information on the project methodology and techniques and had also facilitated the inclusion of labour in the market and it had respected community culture. In addition, it prepared students, as future designers, to engage in producing designs focused on people, ecology and ethics, thus designing in an ethically acceptable manner without losing sight of the consumer market.

Ecological and socially correct initiatives have become significant marketing differentials for any company in any sector. Previously, *Design Possível* partners were universities and non-profit organisations. Today, however, they are also private companies who understand that, with the articulation of the university and the third sector, they can make their ideal of social responsibility effective and efficient. Everyone in the process collaborates for sustainable development actions. The benefit is socialised and the company profits from the social relations and its image in the market. The community and NGOs profit from the effectiveness of their work as they generate income for themselves and achieve social change. The university becomes an articulator of this process, building and applying knowledge at the same pace it trains skilled professionals for a new reality that needs to be created.

The relationship model developed by *Design Possível*, as well as other productive organisation models, seeks to manufacture products and provide visibility of structures and relations constituted in the design and production of those products. In short, the new generation of products and services under development must carry positive attributes, values and quality standards consistent with the pattern of a sustainable society. In the words of the Brundtland Report that defined sustainable development, it must meet today's needs without compromising the natural resources for future generations. Consumers who are sensitive to products attributes not only make use of their functional or aesthetic value but also of their history and their linkages.

If the current economic system, based on an irrational exploitation of nature's resources in order to maintain an exaggerated consumption, is rctained, society as we know it will not survive long. Hence, there is a need for urgent changes. The model of sustainability which underpins *Design Possível* activities is a modest attempt to engage in change. It aims to promote development based not only on the economy but also related to ecology in all its forms, to cultural anthropology and political science. This means following the requirements of eco-development proposed by Ignacy Sachs (2007). He states that eco-development means, in essence, helping the people involved organise, educate themselves, reflect on their problems and identify their needs and potential resources to conceive and design a future that is worth living according to the postulates of social justice and ecological prudence (Sachs, 2007).

The post-industrial world is requiring designers to unite with socially responsible company sectors. Initially, this could be confused with philanthropy, meaning the existence of a joint articulation of the company and community, showing their involvement and responsibilities. The ultimate goal is social welfare, no longer in such a purely paternalistic way as in the philanthropic period, but in order to promote the whole development of human beings. Although the companies present their actions regarding social responsibility, the concept is still closely related to business performance and profitability. This binomial is the key to products focused on consumers' wishes. However, it carries in its core a commitment to shared responsibilities related to sustainability. In most large Brazilian companies that relationship proclaims social and environmental concerns.

In Brazil, since 2005, the ABNT (Associação Brasileira de Normas Técnicas: Brazilian Association of Technical Rules) has enacted the formulation and implementation of policies and objectives that cover legal requirements, ethical commitments, transparency, citizenship and the sustainable development of the organisations' activities. According to Carlos Amorim, ABNT Director of External Relations, Brazil is one of the ten members of ISO, who decided to develop a specific standard on this matter based on international standards (http://www.ethos.org.br/Uniethos/Documents/TextoDSeRSISO26000Tarcilae Celso. Pdf) (Accessed January 28, 2008).

Consequently, a new link of the third sector with the market is under development with private companies as well. They now build a new model of social relations based on entrepreneurship and joint income generation leading to the incorporation of new fair trade principles. The third sector becomes professional, increasingly looking to link with the second sector which will allow for the finding of solutions in areas where the articulation with the State has proved inadequate in tackling certain problems.

In Brazil, over the past five years, productive communities that serve businesses and corporations by providing them with artifacts, products and gifts have multiplied. With their work, they also promote knowledge and income growth, providing worthy conditions to communities and groups that were not assisted or were in a vulnerable situation. To compete with other market sectors these communities resort to an important tool for differentiating their products, management and applied design production. In their circumstances, with the help of *Design Possível*, the producing communities have been able to put a high value on cultural and social characteristics and improve their productive response. It should be noted that their differential is the use of waste produced by the group or society with non-industrial technology capable of reaching a qualified, demanding and profitable market.

Today, more and more people are choosing environmentally friendly and socially committed products. It is the market and the right marketing strategies that prevail. Every consumer seeks to make a difference for the planet when purchasing anything produced by environmentally friendly companies. *Design Possível* ended in 2008 when it was involved in a considerable number of initiatives linked to the third sector.

Several partner institutions (Associação Comunitária Despertar, ONG Projeto Arrastão, Centro Comunitário Irwin Miller, Associação Monte Azul, ONG Aldeia do Futuro, among others) have been assisted by the project. More than 2,500 people have benefited from some type of training and income generation. Even more impressive is the number of price quotations and purchase orders received for the development of exclusive business or proposals from institutions. They actually ended the year with the production and marketing of over 5000 products such as furniture, stationery and personal and home accessories. The follow up of the project figures during 2008 demonstrated the consolidation and growth of productive relationship policies with the third sector by participating companies and institutes.

One final and possibly the most important model involves effective partnership between businesses and third sector agencies. With that partnership, companies develop or create new demands or internal dynamics that, supported by funding and social projects, create an environment of inclusion that makes the insertion of new groups or new productive communities easier. This is the case of Itau Cultural which in 2008 found in the "Despertar" NGO, in partnership with the Gerassol group, the supply of products that would be used in their promotions' campaign.

Also during 2008, members of *Design Possível* looked for the best way to fit the conversion of an informal partnership into a legal entity. Different possibilities were evaluated and several possible arrangements discussed until they came to the model launched in January 2009. The *Design Possível* Project is a partner of the Instituto Papel Solidário (Solidarity Paper Institute). This Institute is located in São Paulo and has worked on special projects for environmental education since 2006. Today it is internationally recognised as a non-profit organisation with the mission of facilitating the reflection, legalisation and management process of solidarity enterprises, non-profit associations and social enterprises (http://www.cidadedemocratica.com.br/perfil/223) (Accessed May 6, 2009).

Advised by the Instituto do Papel Solidário, *Design Possível* began its institutionalisation as a legal entity, as previously mentioned, using the name of Associação Design Possível, a non-profit organisation named after its original project. Its bylaws state that its goal is to promote, encourage, develop and implement design as a promoting factor and a social equity generator. It became clear that this legal development brought new responsibilities and challenges.

With the implementation of the extension project for an NGO, *Design Possível* opened a new market for designers. It has shown that it is possible to work with the third sector and to work towards social development in a sustainable and constant way, consolidating a professional design field for social benefit. The university extension activity gives rise to the professional work, and volunteering becomes a profession, enabling further work and research towards more solid and lasting results.

In this consolidation process, the need for structural alignments arose and the mission of *Design Possível* became more evident. As from that point, it has been split into four departments: pedagogical, commercial, administrative and communication, aiming to systematise the work, train staff and build on its foundations (social development, design research and sustainability). So divided, the departments and their respective coordinators can meet their own needs in a dynamic way. Three departments must build their sustainability on the foundations of their work managed through a cost centre system. They must balance paid and volunteer activities, paying special attention to the third sector or to the social fields.

As a way of disseminating the projects developed by *Design Possível* and to emancipate them, a social strategy of multiplication through a network named Possíveis Multiplicadores has been set up. The goal of this network is to show the possibilities that had already been developed in other Brazilian states. Social technology seeks to meet the human dimension of development and collective interests, ensuring a better quality of life in a sustainable way. As stated by Jacques de Oliva Pena, president of Fundação Banco do Brasil (Bank of Brazil Foundation), there is a large number of social problems to be solved, therefore isolated government, business or civil society organisations initiatives do not meet the demands (http: //www.oei.es/salactsi/tecnologiasocial.pdf) (Accessed: November 12, 2009).

Contributions of the *Design Possível* Network

The economic transformations in the globalised world and the new dimension of ecological awareness bring to discussion socio-environmental responsibility and cause social and ecological actions to make a difference in the market. But, many of the changes that have occurred are still isolated and have little effect. However, it has been confirmed that a socially engaged management is effective in solving concrete problems resulting in social advancement and citizenship training.

The creation of Social Technology Network has established a new paradigm for the development of Brazilian society. Thinking of this structure the *Design Possível* network was founded in 2009. The Universidade Presbiteriana Mackenzie organised the first qualification of possible multipliers with the participation of design students from several states in order to share the methodology developed by the *Design Possível* project.

Students and representatives from NGOs and the community discussed issues related to the development of objects and to the productive management of product design. The goal of this event was to raise awareness of participants of the need to develop similar projects in their areas. With this Possíveis Multiplicadores initiative, *Design Possível* won the Fundação Banco do Brasil de Tecnologia Social award in 2009. This was awarded for the proposal to spread the use of social technology based on the multiplication of social solutions and contributions to the promotion of sustainable development.

> Putting forward design development for sustainability means, therefore, promoting the capacity of the production system to respond to the pursuit of social welfare, using a number of environmental resources dramatically below the levels that currently prevail (Manzini and Vezzoli 2003: 23).

Members of the *Design Possível* network realised that to promote, encourage, discuss, study and implement design as a tool for social transformation, it should be present in the student movements, which in the professional field of design are represented by the National Board of Design Students (CONE). This is a fertile ground for actions that seek to raise doubts about current design models and this challenge may create players who are able to lead design in a whole new direction.

A 40-hours event was arranged in São Paulo to permit the exchange of experience, the making of presentations and the organisation of activities all focused on the work of *Design Possível*. Attention was also given to the construction of a collaborative methodology that could be used in other states where it was intended to implement the *Design Possível* network. Currently, *Design Possível* is active in Amazonas, Pará, Bahia, Rio de Janeiro, Paraná and Santa Catarina.

Workshops, group dynamics and lectures were also presented by the *Design Possível* group. Social responsibility projects in partner communities were visited. The purpose of this integration was to provide a collaborative approach to teaching and to encourage new members who were seeking local development and integration in different regions of the country to join the *Design Possível* project.

It should be added that this work disclosed the methodology of product design used by *Design Possível* with the purpose of generating income and creating opportunities for educational growth in areas diagnosed as highly vulnerable, where the Human Development Index was low and the social structure insufficient to meet the needs of the population. It was possible to determine that the project members and its multipliers built a knowledge repertoire, acquired experience and positive expectations regarding the reapplication of the working methodology and management experience gained during the five days of the event. It was noticed that changes were also incorporated in the creation of objects, taking into consideration the environmental and socio-economic issues of business sustainability in environmental terms.

After the initial activities, participants continued their training via distance learning for six months to learn how to apply their acquired knowledge in their local communities. The work of *Design Possível* multipliers in productive communities has recognised the value of these communities' cultural and social characteristics. Their productive response has improved, optimising the use of waste produced by the group itself or the society. Therefore, using discarded materials it has been possible to develop products with significant added value, thus reaching the qualified, demanding and lucrative market, more consistent with a society pattern that wants to be built through the sustainable development parameters.

Conclusion

Regarding the *Design Possível* project development at Universidade Presbiteriana Mackenzie (Brazil) and Università degli Studi di Firenze (Italy), it was noticed that in the initial phase the interaction was complex due to communication difficulties between Brazilian and Italian students. Besides these initial problems, Italian students suggested many productions.

Another mismatch occurred as Brazilian students solved problems, presenting mostly practical, simple but not innovative solutions. Meanwhile, Italian students looked for new concepts regardless of their realisation. They sought to create something new without taking into account the cultural and social realities of Brazilian NGOs. Clearly the participants came from two different backgrounds, one highly technical and the other theoretical. On the other hand, throughout the project development, the union of two different backgrounds from the universities proved very productive since the action models of the two countries complemented each other. Cooperation is based precisely on the discovery of complementarity.

Design Possível sought the cooperation of universities from different countries, creating a relationship aimed at helping poor communities associated with Brazilian NGOs. This provided students and teachers with a learning process to complement their design background. In this sense, the project attempted to find different uses for design, opening a new perspective in search of societal transformation. Is was not a project focused only on industrial production, it also focused on handicraft production and on increasing professional activities in poor communities excluded from both consumer and producer markets.

Numerous activities were carried out and actions taken by the project including: workshops for product development; technical visits to NGOs; monitoring of manufacturing prototypes; discussions about products; and dialogues about design and the current environmental issues. Those actions have undoubtedly contributed to the process of teaching and learning and to the personal growth of participants in both universities as well as in the NGOs.

We believe that universities have the mission of developing the students' potential and encouraging them to think and reflect on the importance of design and from that to find shared solutions, generating new life attitudes, in respect of natural resources and social and economic development. With regards to the *Design Possível* NGO, it was noticed that the knowledge acquired from the project started at Universidade Presbiteriana Mackenzie and Universita di Firenzi was very effective in strengthening and encouraging the application of design as a generator of social equity. Also, it showed that one can work alongside the third sector and private companies toward the development of people with environmental responsibility.

The newly created *Design Possível* network has been investing in the training of its multipliers. It aims to provide subsidies in collaboration with design students to implement new social responsibility projects in poor regions in other Brazilian states. The goal is to disseminate and implement the use of social technology. Once put into practice, it will promote, encourage, discuss and implement design as a tool for social transformation, considering the project proposal and the social technology strategy employed.

Taking into account these two factors – the project proposal and social technology – a model of community development and inclusion will be created, showing social, environmental and economic results that are infrequently valued. It is expected that at the end of the whole development process, a new model of social relationships among managers, communities and designers will be consolidated. A designer can facilitate relations and social changes, using its close views of the project as the sole tool. He or she needs to be a multiplier of sustainable and ethical practices, and also he or she should encourage the community to reflect on the search for new attitudes for a better future.

References

Brazil. 1999. *Law No. 9795*, April 27, 1999.

Fundação Banco do Brasil. 2009. *Tecnologia social uma estratégia para o desenvolvimento.* http://www.oei.es/salactsi/Tecnologiasocial.pdf (accessed November 12, 2009).

ICSID. 2010. *International council of societies of industrial design.* http://www.icsid.org/about/about/articles.31 (accessed March 9, 2010).

Manzini, E, and C O Vezzoli. 2003. *Desenvolvimento de Produtos Sustentáveis.* São Paulo: EDUSP.

Papanek, V. 1995. *Arquitectura e Design: Ecologia e Ética.* Lisbon: Edition 70.

Philippi Jr., A, and M C F Pelicioni. 2000. *Educação Ambiental: Desenvolvimento de Cursos e Projetos.* São Paulo: Signus.

Pons, I. E. R. 2006. *Design Possível Um Estudo de Caso Exploratório de Praticas Educativas Desenvolvidas com ONGs (2004–2005).* Master's diss., Postgraduate Programme in Education, Art and Cultural History at the Universidade Presbiteriana Mackenzie.

Sachs, I. 2007. *Rumo à Ecossocioeconomia – Teoria e Prática do Desenvolvimento.* São Paulo: Cortez.

Sorrentino, M, R Trajber, and L A Ferraro Jr. 2005. Educação Ambiental como política pública. São Paulo. *Education and Research Journal of USP* 31(2): 287. May/August.

Chapter 12
Education for Sustainability, Ethical Relations

Eliete Jussara Nogueira

Introduction

Environmental issues have emerged in recent decades as part of a civilisation crisis. There is growing concern from different social sectors regarding the negative impacts of technological and economic hegemony. The accumulation of capital with short term profit goals, population growth and the limited resources of the planet are reflected in an environment in which the quality of human life is at risk. We can see such tragic effects in: the pesticides used in food; the loss of biodiversity; the extinction of species; the constant fear of robberies and violence in general in the metropolis; the excessive garbage with no place for it to be discarded; and children who do not play in the streets anymore and are always secluded in schools or other institutions for their personal security. These are just a few examples drawn from many everyday situations that comprise human relationships as we know them.

The advent of the industrial revolution, focusing on increased production, the modification of consumption and work relations and profit at any cost, has increased the environmental crisis. The industrial logic, being productivist, utilitarian and consumerist, has gradually created an extreme situation in which nature is perceived as a reality for humans to exploit to their advantage. A change in the concept of nature emerges from the perception that humans are inserted into it. Considering the historical influences of social production, Serres (1990) states that nature is human nature itself. Therefore, disregarding the time of natural processes, humans have compromised their own human lives, conducting their organisms through the rhythm set by the clock. Typical of the modern world, the advent of industry and technology has pulled humans away from nature's plan, from the rational animal, with the mechanisation of work; they

E.J. Nogueira (✉)
University of Sorocaba (UNISO/SP), Sorocaba, SP, Brazil
e-mail: eliete.nogueira@prof.uniso.br

M.L. de Amorim Soares and L. Petarnella (eds.), *Schooling for Sustainable Development in South America*, Schooling for Sustainable Development 2,
DOI 10.1007/978-94-007-1754-1_12,

became the machine people, who were converted into a work force that functioned in marked and disciplined time, working to the rhythms of the cogs of production.

Working time converges into an economy category in the capitalist system. Salary is measured by the cost, by the value assigned to the work force, materialised in money to obtain consumption goods. We witness another transformation of humans, from producers to consumers. In this century we live in a society that encourages consumerism (Bauman 2009). In a sociological analysis of the contemporary world, Bauman defines its main feature as liquidity, identifying that we are in liquid times, liquid human relationships, liquid fear, thus modernity is liquid. The author uses the 'liquid' metaphor to emphasise the ephemeral character of the world following a more 'concrete' period of modernity defined as the age of reason, of illuminism, of science, of theoretical certainties, of metanarratives. The move was to a world full of contradictions and uncertainties. Gradually society is being built with the purpose of living for consumerism. Work relations have been transformed into relations of purchase and sale, including in this relationship of consumption the human being himself or herself, who is also something that can be consumed. There is an excessive concern with appearance (packaging), with remaining young (new), in qualifying or describing skills that can make the person a faster selling product (Bauman 2008, 2009).

The insertion of society into a cycle of global consumption has generated an increase of products that are not essential for life resulting in an increase in waste. The act of consuming in the contemporary world gains a subjective character. Consumption is no longer dictated by necessity but by desire, by will, by the urge to buy, since in this post-industrial Western culture human happiness is associated with the ability to purchase and to accumulate assets. Buying compulsively, without need, even if the items purchased are quickly discarded or not used at all, is a behaviour encouraged by the media in general and it is facilitated by payment in installments. Consequently humans are not only compulsive, they are also indebted (Bauman 2009). Economically, consumerism maintains the relationships between capital and profit.

The consumerist relationship is linked to desires, subjective sensations, feelings of emptiness, in which according to this logic these can only be fulfilled by buying something new, presented by the market as the solution that must be consumed in no time. However, human feelings are complex and subjective; they are not subjected to the laws of the market and cannot be satisfied by things. In that context contemporary humans will always be insatiable and unhappy. In a society that encourages consumerism, everybody is a target for the strategies to transform human beings into consumers,

> As soon as they learn to read, or perhaps long before that, "shopping dependence" is established in children. There are no distinct training strategies for boys and girls – the consumer's role, unlike the producer's, has no specificity in gender. In a consumer's society, everyone needs to be, must be and has to be a consumer by vocation (in other words, to see and treat consumption as a vocation). In this society, consumption seen and treated as vocation is both a right and a universal human duty that knows no exception (Bauman 2008: 73).

The principle 'time is money' demonstrates the social representation of the hierarchy of values in contemporary life. In this priority, human beings overestimate the things that can be bought, and push to one side the importance of feelings of solidarity, affection and pleasure in social relationships. Thus, people do not perceive the artificial value of things and they impose a hasty rhythm of life which may result in psychological problems such as phobias, disorders and stress. The hurried pace of life triggers faster and faster responses, enhancing the chances of behaving without thinking about long-term consequences. The environmental imbalance, according to Mazzotti (1997), is the result of this modern consumerist life style. The base for a sustainable society should take into account ethics that values respect for cultural biological diversity.

This consumerism's modern liquid phenomenon designates a hasty life where the instant matters. The need to consume, discard and replace, promotes a life with no ties, no history. A consumerist economy is based on excess and waste. The idea of instant and individual happiness is present in everyday life, in books, in messages and images of how much we need to invest in ourselves, or to do it yourself, or even to take care of yourself, and take time for yourself. These increase the individualistic idea of a consumption society which produces only more consumption. However, the human being as a social being has always liked to live in groups, in society. However, in liquid modernity, the other must be banished; he/she is dangerous to your individual freedom. But, without the supportive affection of others freedom or a sense of freedom is meaningless. It does not cause emotion and does not promote happiness (Bauman 2009).

For the survival of humanity, Gatarri (1990) proposes an ethical-political articulation, which he calls ecosophy that comprehends the environment, social relations and human subjectivity. Changes in the way of thinking and acting do not always occur naturally, therefore there is a need for ethical-political articulation to build the rules and laws that will favour attention to urgent environmental issues by society,

> The same ethical-political perspective runs through the issues of racism, phallocentrism, the legacy of disaster by a so-called modern urbanism, the artistic creation set free from the market system, the pedagogy capable of inventing its social mediators, etc. This is the issue, after all, of the production of human existence in new historical contexts (Gatarri 1990: 15).

Different paths can be shown as possibilities to minimise the alarming levels of degradation that nature has been suffering. Everyone must go through changes in behaviour, values and attitudes towards everyday life. Recommendations resulting from encounters and international meetings indicate intentions to change the mentality and actions regarding environmental problems. Education can be one of the forms to implement relationships between people and nature in a sustainable way. If "all knowledge is self-knowledge" (Santos 2003: 92), the knowledge applied to nature is nothing but the knowledge humans have about themselves, so that as we educate the ethical person we also build environmental awareness.

Environmental education is a permanent awareness process about knowledge, individual and collective values and attitudes regarding environmental problems through an interdisciplinary focus. This definition has been adopted in Brazil and in most countries that are part of the UN and has as its goal to redirect education to sustainable development. Pelicioni (1998) highlights the responsibility taken in the Global Forum of Eco–92 with the principles: environmental education as a right for all; a foundation for critical and innovative thinking; a local and global ideological conscience; and a way of integrating knowledge, opinions, values and educational actions for sustainability. Changing the hegemonic logic and turning to teaching for understanding the changing world should initiate actions that allow the building of critical knowledge to face environmental problems created by capitalism. Thus, environmental education is undertaken as a political action, a social practice linked to sensitisation and reflection on economic, political, daily issues, seeking to break the school-forest dichotomy (Reigota 2002).

Manzochi (2009) proposes a theoretical guideline for an environmental education that forms citizenship in an emancipatory perspective. As such, it should work on the dimensions of concepts, values, and participation in political actions. In this context, we must consider what an environmentally educated citizen would be. Manzochi suggests that an environmentally educated person would be one capable of producing/participating in changes related to environmental issues in the world in which he/she lives, being able to understand environmental problems, to place himself/herself in the social group, and to have the effective capacity of acting politically on these issues. Educational processes must converge to help people to understand political, structural, and organisational determinants of the problems that affect the environment, focusing on interventions that lead to changes in society.

The school is a privileged space for the development of behaviours, values and attitudes towards social life. Many individuals attend school for a considerable period of time; therefore we can infer that this is an important institution for the upbringing of people. The school witnesses the whole human development from childhood to adulthood and interferes with the basic structures of subjectivity mainly built in children.

In general, psychology in a socio-historical perspective understands the constitution of the subject determined by the interaction of ontogeny, phylogeny and sociogenesis, that is, the history of life, species, and society to which the subject belongs. Vygotsky (1998) suggested the formation of higher psychological functions, thinking, imagining, creating, as determined by the formation of signs, self-generated stimulation that acts on the individual. The signs, among them language, have great importance in the individual's way of thinking and they are formed and determined by society. According to Vygotsky, the construction of the signs by the child is mediated by the society with which he or she interacts and the school is an institution built intentionally for this interaction.

Thus, we can relate the mediation of the school with the experiences that build the child's subjectivity. Such mediations have been studied by different authors showing the ideological action of the school. The organisational structure in school rules, in the location of the desks, in the times of lessons and the choices of curriculum content, project the values of an authoritarian system. Foucault examined modern structures aimed at disciplining people and he points out that in institutions of constant vigilance, such as schools, hospitals and prisons. "… it is not only the prisoners who are treated as children. But also children as prisoners. Children suffer an infantilisation that it is not from them. In this sense, schools are a bit like prisons" (Foucault 1977: 42).

According to Gallo (2009), one way that schools overcome this is the constitution of singular and autonomous subjectivities that do not submit to the values of the production machine ideology. The concern of school actions should focus much less on the content area and more on the methodologies used for teaching. This is because what matters is not the information that is taught but how it is taught since values are passed through attitudes and through the ways school relationships are established among teachers, students, staff and the community.

The authoritarian relationship between teachers and students, the monitored controlled environment, the schools being more and more enclosed by high walls, inject into subjective structures respect for authority, for superior power, the action of fearing the other. Regardless of the content being taught, learning is obedience to power, to the imposed rules. The everyday actions of the school shape a way of perceiving the world and relating to it. Thinking about the ethics embracing school relations is really thinking about the subjective formation of individuals. When we internalise ethical signs, our way of thinking is guided by collective environmental interests that influence directly our action towards the other.

A subversive attitude which may question the hegemonic ideological action must develop autonomous structures, creating singular and free individuals. Using philosophical concepts, Gallo (2003) compares the larger education, the policies, the plans, parameters and national guidelines thought of and produced for the dominant power, as control machines. This larger education produced in macro policy has as its premise that education must correspond to learning. However, learning is out of control. Even when we have lesson plans, teaching methodologies and evaluations, we have no guarantee that learning occurs as planned. Consequently, also according to Gallo, the classroom ambit, the micro policy, expresses everyday actions which correspond to war machines with a profile of militant teachers awakening desires. This smaller education leads us to new encounters and escapes, necessarily with political content, challenging the prevailing hegemony.

Another way of overcoming hegemonic ideology is experience through contradiction. The school as a territory of contradictions can creatively implement pedagogical practices that invite students to transgress the hegemonic order. Soares (2001) presents some possibilities of emancipatory pedagogical practices offering creative ways to build uncontrolled knowledge. These include proposals of living art, poetry in school daily life, with the concept of transforming the values of a world that makes people develop a mental stereotype about something by using the mass media. Such practices suggest that the transformation of daily life comes from the student and that may widen fissures, pass through the gaps, find breaks, win the alienating daily life through cultural revolution, through the sense of desire.

In the context of the pedagogy of the senses, in the proposal of knowledge as nomadic (knowledge as always roaming about, not finished, in movement) and rhizomic (a horizontal knowledge which multiplies the interactions that are originated in it), "the knowledge becomes flavours", children want what they do, it is an invitation to knowledge, and not an imposition "...under an ethical and aesthetic view, beyond good and evil, an artistic and creative reality, free of divine imaginary, of judgment, of the truth, of the punishment and correction" (Lins 2005: 1230).

Through the ethics of affection rather than fear, it may develop an intelligence of the sensitive and the desirable which is the inner force propelling transformations.

The proposal for a school as a space for pedagogy of the senses can deviate students from a predetermined destiny. Considering that for Deleuze (1993: 166), "Learning is only the intermediary between non-knowledge and knowledge, the living passage fron one to the other", the school should be more concerned about the experiences it provides to the apprentice,

> The apprentice, on the other hand, raises each faculty to the level of its transcendent exercise. With regard to sensibility, he attempts to give birth to that second power which grasps that which can only be sensed. This is the education of the senses. From one faculty to another is communicated a violence which nevertheless always understands the Other through the perfection of each. On the basis of which signs within sensibility, by which treasures of memory, under torsions determined by the singularities of which Idea will thought be aroused? We never know in advance how someone will learn: by means of what loves someone becomes good at Latin, what encounters make them a philosopher, or in what dictionaries they learn to think (Deleuze 1993: 165)

Ethical Relationships in Schools

When thinking about changing attitudes towards environmental problems or other problems in today's world, we must understand the ways of society interactions. Understanding how working and social relationships are structured is necessary to reflect about ethical and moral issues. Philosophy distinguishes the terms 'moral' and 'ethics'. Sense and moral conscience are used with regard to the values, feelings, attitudes, the relationships we have with other people, including the evaluation we make about good and evil and the desire for happiness. From this understanding of morality, every society would then have a set of values according to its own culture, which would ensure the physical and psychological integrity of its members. The field of ethics has as an imperative condition the conscience and moral responsibility of individuals in permanent reflection about the consequences for themselves and others, consisting of discussion, evaluation and interpretation of the meaning of moral values (Chauí 2001).

The existence of moral values does not mean the explicit presence of ethics; the ethical subject would be the one who "knows what he/she does, knows its causes and the purposes of his/her actions, the meaning and his/her attitudes, and the essence of moral values" (Chauí 2001: 341). The will of ethical action cannot be subordinated to the will of others, to personal interests, or to the market. Therefore, one of the conditions for the constitution of ethical human beings is freedom, to be free for self-determination, to act ethically considering what is best for all, and not by fear of punishment which would make them slaves.

The questions regarding the planet's sustainability and the pedagogical practices developed in schools go through the understanding of which ethics the educator is valuing and defining in his/her actions. For Costa (2009), to distinguish environmental ethics from ecological ethics seems fundamental for the development of a new perspective on environmental issues. From an environmental perspective, according to Costa, anthropocentrism becomes evident. The thinking person is the explorer of nature seeking his or her needs and desires, the only creature among other living creatures with full rights to occupy the planet. Putting the human being at the centre subjugates other species and it does not recognise the importance of everyone in an integrated system of life. It values merely utilitarian interests, depending on the market without limits.

As opposed to this environmental ethics, we look for ecological ethics where human beings share their lives with other species, being aware that their existence is as important as the others on the planet, neither better nor worse. And, therefore, the integrity of the ecological community should be preserved. In this argument there is an intrinsic value of life which attributes respect and moral value to all relationships that ecologically contribute to life on the planet in an integrative conception.

People are able to reflect on fair values of collective interest when they interact with situations in which they are required to make choices and reflect on moral issues through dialogue. In approach, moral formation passes through reciprocal relationships among students, teachers, members of the community, in the active exercise of mutual respect and in the exchange of views, since human beings are relational and they are constituted by relationships with other human beings during their life history.

The school in our society acts to form citizens; it has an historical commitment to solidarity as a way to establish ethics. In the development of his autonomy pedagogy, Paulo Freire (1997) defended a pedagogical practice in favour of pupils' autonomy. And, for him, that demands the acceptance of newness, the rejection of any form of discrimination and prejudice, the critical reflection on their own educational practice, the recognition of cultural identity incorporating diversity, and the understanding of reality as presented. This occurs without forgetting the conviction that change is possible and that education is a way to intervene in the conscious world so that the interests of the market do not overlap those of humanity. For Freire, the principle that the process of education demands freedom should be an advocated principle for the development of children's autonomy in school daily life.

For a better understanding of the development of morality in children, Piaget (1977) demonstrates in his researches that the child goes from a state of anomie (lack of rules) in early childhood to heteronomy (rules established by external factors), and in the third stage of morality, autonomy (where the rules are established by internal factors). In the final stage, the human being is able to self govern, to analyse the context and to make choices for everyone's well being. Piaget also states that the way of reaching autonomy depends on how rules are established for the child. The imposition of values or rules may lead children to fulfill them through fear of the consequences or else they develop other ways of dealing with situations of social conflict, such as lying or running away, in essence not assuming responsibility for their actions. In contrast to that, Piaget defends freedom of thinking, the adoption of some rules by the children through the understanding of how much these rules are good for them and for others.

During the years of military government in Brazil from 1969 to 1986 educational policies intended to strengthen national unity cultivated in the school environment heroes and symbols of the mother country to maintain order. In a more incisive way, specific subjects were implemented in schools: Moral and Civic Education and Studies of Brazilian Problems were both based on the belief that moral values are universal and can be learned through receiving mere information about the rules. The moral education advocated and the Brazilian problems presented at the time were previously determined; both the choice of problems and their solutions. The way of viewing them was indoctrination, the students were subordinated to hegemonic thinking that favoured the governors. In the context of dictatorship, this condition of teaching and learning of imposed contents resulted in years of silence for teachers and for pupils the behaviour of obedience to rules even without understanding them. Nowadays, in a democratic regime, the National Curricular Parameters (1998) provide guidelines for teaching practices and they include ethics as a cross-curricular theme, understanding that it must be taught in all disciplines and be included in school relationships as an integrated whole. The proposal is based on Piaget's concepts regarding moral development through building values in the interactions between the subject and object of knowledge but it still maintains the hegemonic and consumerist speech of the market. In two distinct governmental periods the development of moral values through the school has continued to be recognised as a way of changing social attitudes.

Moral teaching is not possible with sermons, speeches or specific disciplines; a school environment with mutual respect practised and cultivated in an atmosphere of cooperation can contribute much to the achievement of an autonomous morality. In this context, the school should provide situations for reflection and also awareness of consequences and responsibilities that our actions generate in order to develop moral attitudes in a free environment.

The school presents itself as an area of individualism and there is a loss of unity when it fragments timetables and curricula, separates or classifies students in the classroom, or in everyday situations reinforces individualistic behaviours to the detriment of cooperative ones, for example, when the teacher asks the students to take care of their own things, not to look in the other student's notebook, that they must be quiet at their desks, and that if someone gets hurt or fights they should call the teacher. It centralises power and takes away the autonomy and the creative possibility of building knowledge. The dialogical relationship that is so necessary for ethical experiences is abandoned in exchange for the bureaucratic fulfillment of contents. The ethical relationship, or the development of attitudes socially recognised as moral, is not directly linked to intellectual development; it is located in the experience of interactions with other human beings that facilitate significant learning. Humans have always learned from other humans. The human being is constituted by the presence of another, that is, we form our identity in social mediation, our thinking is socially and historically determined, so that all knowledge is cultural and related to the historical moment in which it was built (Vygotsky 1998).

Thinking of pedagogical practice in the contemporary context that responds to the serious problems we live with, such as drugs, violence, ecological degradation, genetic intervention, among other uncontrollable advances that endanger life on the planet, means reviewing ethics based on rational foundations and pursuing the development of an intellectual and social autonomy of the subjects involved in the educational process. Beyond the contents, a new ethic needs to be present in the relationships in school. For Santos (2003), the expansion of democracy occurs in overcoming the authoritarianism that exists in social relationships in every structural space of social interactions, among them the school. We must, then, recognise the political character of social relationships existing in the school and transform them from power relationships into relationships of shared authority.

The discussions about the rights and duties of citizens in a vertical view can be magnified in school to revalue the principles of equality among different forms of knowledge. Examples in school daily life include the distinction between capitalist nature and anti-capitalist ecology, or hegemonic models of democracy and participatory democracy. There are possibilities for experiences in school that should emerge from dialogue and conflict in the production of new relationships in the school's daily life. Since every code of ethics is ambivalent, containing contradictions and conflicts, we must accept that humans are imperfect and ethical behaviour always depends on the individual decision of the will and desires of these humans facing concrete circumstances. The challenge for the school is to find ways of developing subjective identities with principles that work as a basis for moral conduct.

Conclusion

The dry tap
(But worse: the lack of thirst)
the lights off
(But worse: the taste of the dark)
the closed door
(But worse: the key inside)

José Paulo Paes (2000: 137)

The economic models adopted throughout history have generated strong concentrations of income and social inequalities. They have degraded the quality of life and health of people. At the same time, they have damaged the environment through the predatory exploitation of natural resources and pollution. We are entitled to an ecologically balanced environment and it is the role of government and the whole community to stand against the values of the market to preserve it. It seems to be necessary to establish limits and rules that will organise the interventions of humans in nature.

The practical and utilitarian concept of economic systems has transformed nature and people into production factors. In this approach, nature, a social asset and everyone's right, is seen and treated as a business belonging to a few people. Those privileged ones can explore it using market laws independently of human issues. Calculating costs and profits, they regard nature and people as things, disposable and manipulated objects in favour of consumerism.

Education is a recognised way of providing an understanding and providing opportunities for seeking solutions to complex and diverse problems related to changes brought about by human activity. Specifically, the school needs a new order that incorporates respect for the complexity of environmental issues with an ethical and interdisciplinary approach committed to social practice and dialogic interaction. Personal and collective paths should be discussed, planned and taken to appropriate actions for life on the planet. The school needs to be challenged to change the logic of knowledge building and implement a cooperative culture in order to stimulate creativity and the capacity for transformation compatible with sustainability of life on earth.

With globalization, we have become more worried about producing and consuming than with the consequences of these for life on the planet. The worst affected by the greed of the privileged continue to be the poor who increasingly need money to purchase goods for their subsistence, since the land no longer belongs to them. It is necessary to stop, to think, to invent, and to reduce the levels of consumption and production, at least until solutions for a sustainable world can be found. We are capable of redistributing food, making renewable energy work, cleaning the air we breathe, co-existing with other living creatures and sustaining our quality of life, provided that appropriate political and individual decisions are made as priorities.

The school needs to go back to questioning, discussing multiple knowledges, articulating them in its human relationships and in the social and cultural contributions it brings to the world. The practice that reveals creative ways directed to the senses allows socially relevant knowledges. It understands the educational act as an intersubjective construction, woven into networks of interactions with the other, that was very much forgotten in an individualised society.

Working with a new ethic means leaving sameness and common sense, looking for different dimensions of the hegemonic social model, therefore, losing status and social distinctions, to be neither rich nor poor, neither arrogant nor humble, neither powerful nor disempowered. It also means moving forward in the understanding of what is and what can represent the school daily life as emancipatory educational practices, instead of thinking of the other as different from yourself, thinking of him/her as the other within yourself, as the possibility of being human.

Sustainability, a theme that today presents itself as a multidimensional problem and which demands responsible reflection with regard to future generations and also demands that schools educate morally autonomous people. A change in mentality needs to be stimulated in the educational process in order to educate individuals who are capable of critically analysing political attitudes, values, rules, and laws that determine collective social life with respect for an interconnected environment necessary for life on the planet. Thinking about ethics in school relations is complex, because, in one way or another, each one has his or her own ethical convictions, influenced by religion, by his or her life history, by his or her experiences, which make up his or her uniqueness, but a collective stance on the degradation of the earth's ecosystem is urgent, it is essential to establish limits that preserve life.

The metaphor of the hunter and the gardener used by Bauman (2009) differentiates the way that people are in the world. The hunter in a society where there is nothing else to hunt can be scary since the hunter's behaviour is guided by defence and attack. The gardener is governed by another ethic, the one where the world depends on constant attention and effort from each one. Increasing the power of utopias of a sustainable world requires, among other things, confidence in human potential. It is necessary to believe that the people are capable of realising what is wrong and what needs to be changed, and demystifying the values of the dominant lifestyle which define a path to a superior society and that actually the belief dictated by the market does not allow achieving what was promised – happiness. If the beliefs of the modern era no longer support themselves by the results achieved by them, their strategies must also be criticised.

> Our lives, whether we know it or not, and whether we salute or lament them, are works of art. To live as required by the art of life we must, like any other kind of artist, establish challenges that are (at least at the moment they are established) difficult to confront directly; we must choose targets that are (at least at the moment of choice) far beyond our reach, and standards of excellence that, in a disturbing way, seem to remain stubbornly well above our capacity (at least the reached one) to harmonise with whatever we are or may be doing. We need to *attempt the impossible* (Bauman 2009: 31).

References

Bauman, Z. 2008. *Vida Para Consumo: a Transformação das Pessoas em Mercadoria*.Trans. A. M. Carlos, Rio de Janeiro: Jorge Zahar Ed.

Bauman, Z. 2009. *A Arte da Vida*. Trans. C. Al. Medeiros, Rio de Janeiro: Jorde Zahar Ed.

Brasil, Secretaria de Educação Fundamental. 1998. *Parâmetros Curriculares Nacionais; Terceiro e Quarto Ciclo: Apresentação dos Temas Transversais*. Brasília: MEC/SEF.

Chauí, M. 2001. *Convite à Filosofia*. São Paulo: Ed. Ática.

Costa, C A F. 2009. Ética ecológica e medioambiental? *Acta Amazônica* 39(1): 113–120. Manaus.

de Santos, B S. 2003. *Um Discurso Sobre as Ciências*. São Paulo: Cortez.

de Soares, M L A. 2001. *Girassóis ou Heliantos: Maneiras Criadoras Para o Conhecer Geográfico*. Sorocaba: PM–Linc.

Deleuze, G. 1993. *Difference and repetition*. Trans. Paul Patton. New York: Columbia University Press.

Foucault, M. 1977. *Vigiar e Punir*. Petrópolis: Vozes.

Freire, P. 1997. *Pedagogia da Autonomia: Saberes Necessários à Prática Educativa*. São Paulo: Editora Paz e Terra.

Gallo, S. 2003. *Deleuze e a Educação*. Belo Horizonte: Ed. Autêntica.

Gallo, S. 2009. *Subjetividade, Ideologia e Educação*. Campinas: Editora Alínea.

Gatarri, F. 1990. *As Três Ecologias*. Trans. M. C. f. Bittencurt. Campinas: Papirus.

Lins, D. 2005. Mangue's school ou por uma pedagogia rizomática. *Educação e Sociedade. Campinas* 26(93): 1229–1256.

Manzochi, L. H. 2009. Reflexões sobre o potencial educativo de alguns recursos didáticos do campo teórico-metodológico de "conflito sociambiental" na formação continuada de professores em educação ambiental. *Revista de Estudos Universitários* , dez.Sorocaba, SP. 35(2): 185–208.

Mazzotti, T B. 1997. Representação social de problema ambiental: uma contribuição à educação ambiental. *Revista Brasileira de Estudos Pedagógicos. Brasília* 78(188–190): 86–123.

Paes, J P. 2000. *Os Melhores Poemas de José Paulo Paes. Seleção de Davi Arrigucci Junior*, 3rd ed. São Paulo: Global.

Pelicioni, M C F. 1998. Educação ambiental, qualidade de vida e sustentabilidade. *Saúde e Sociedade* 7(2): 19–31.

Piaget, J. 1977. *O Julgamento Moral na Criança*. São Paulo: Mestre Jou.

Reigota, M. 2002. *A Floresta e a Escola: Por uma Educação Ambiental Pós-Moderna*. São Paulo: Cortez.

Serres, M. 1990. *O Contrato Natural*. São Paulo: Nova Fronteira.

Vygotsky, L S. 1998. *Formação Social da Mente: O Desenvolvimento dos Processos Psicológicos Superiores*. São Paulo: Ed. Martins Fontes.

Chapter 13
Digital Literacy and Sustainability: The *Vozes que Ecoam* Project

Luiz Fernando Gomes

Introduction

If it is true that education and development make an irreducible binomial and that development must be sustainable, then we need an education for sustainability. Since we are discussing education and sustainability with the economy as the background, we must remember that there are the rich and the poor and this economy does not operate on both in the same way. Therefore, the educational model we have chosen has embedded sustainability.

Benfica (undated), an historian from the Federal University of Bahia (Universidade Federal da Bahia: UFBA), defends the above point of view and stresses the fact that the dominant production model is a destructive one. In fact, the author, based on Altvater (1995: 305), states that the term sustainability has already been worn out and has become a "logical absurdity" because development and sustainability are logically incompatible. Trying to counteract and go against this logic, we need an education that seeks and promotes harmonious integration between people and the environment. Gadotti (2000) in his article "Pedagogia da terra e cultura da sustentabilidade" ("Pedagogy of the earth and the sustainability culture") suggests some knowledge and values for this pedagogy. According to Gadotti, the pedagogy of the earth is based on a new ethical paradigm and a new intelligence in the world. An intelligent person is not the one who knows how to solve problems but the one who has a life project of solidarity, given that solidarity is not just a value but a condition of survival for all. The pedagogy of the earth proposed by Gadotti also includes education for simplicity and stillness. The author argues that people need to be guided by new values, such as simplicity, austerity, calm, peace, learning to listen,

L.F. Gomes (✉)
University of Sorocaba (UNISO/SP), Sorocaba, São Paulo, Brazil
e-mail: luiz.gomes@prof.uniso.br

M.L. de Amorim Soares and L. Petarnella (eds.), *Schooling for Sustainable Development in South America*, Schooling for Sustainable Development 2,
DOI 10.1007/978-94-007-1754-1_13,

learning to live together, sharing and discovering and doing together, and finally, people must change their consumption habits by reducing their demands.

Needless to say, at least in Latin American countries, the school curriculum, in most cases, deals with the issue of sustainability focusing on the accumulation of garbage, recycling still incipient in much of the domestic and industrial waste and called environmental conservation, focused on the extinction of animal and plant species and vegetable and mineral resources. The global economy, while promoting 'sustainability', urges a life for consumption exacerbated by consumerism.

The internet and widespread use of the computer cause a conflict between local and global concerns. Many people do not identify with their communities and their interests, each consuming the same goods, the same messages and the same images, as Hall (2005) reminds us when he writes about cultural identity in post-modernity. The uncritical use of technologies leads us to a kind of technological totalitarianism (Armitage 1999). For Armitage, the pan-capitalist neo-liberal discourse legitimises the political control, cultural and ideological people. Against this tendency, Armitage preaches a critique of the neo-liberal model by those who are "disenchanted with the increasing substitution of education for mere technocratic information" promoting among those excluded arguments against active policies and strategies.

The project *Vozes que Ecoam* (Echoing Voices) was undertaken for ten months in 2008 in the suburbs of a town in the countryside of São Paulo state, Brazil. As will be seen below, it sought to offer a counter-discourse, to promote an alternative form of digital inclusion in an out-of-school environment, aimed at rebuilding the identities of the participants and their communities. The activities undertaken did not address 'ecological' or 'sustainable development' issues in their usual sense, but aimed, through a critical pedagogy guided by multiliteracy (especially visual, verbal and digital) to offer an alternative curriculum and a chance to rethink local communities, their needs and desires and to make their voices echo through the technological tools of communication.

We begin by discussing and criticising the sustainability discourse, then opposing it to the pedagogical discourse that guided the project *Vozes que Ecoam*. After presenting the project and its development, we conclude by discussing the effects of the literacy events on the attendees as well as on the communities involved.

The Sustainability Discourse

Poverty and ecological problems are interdependent, because as Sachs (1986) states and reviewed by Lima (1997: 2) in his extensive article on sustainability, "environmental degradation worsens the living conditions of the poor, their poverty leads to a predatory exploitation of natural resources, closing a perverse cycle of social and environmental damage". The current global economic model is guided by an unlimited development proposal from a finite resource base. As Lima (1997) warns, this model is unsustainable in the long run. To justify this claim, Lima presents some of the major criticisms that have been made of this model, such as those of Buarque

(1990) who argued that its technique and ethics do not go hand in hand. Moreover, the market system does not meet the needs of the people. It leads to consumerism, with an eye on profit, generating areas of social inclusion and exclusion. Such an economy based on inequality has political consequences, since "truly democratic proposals hardly can be sustained over very uneven patterns of income distribution" (Lima 1997: 5).

According to Lima, the relationship between development and environment makes us realise that the North of the globe suffers from the pollution from wealth and from industrial waste and the South suffers from pollution, poverty, malnutrition, lack of sanitation, etc. To discuss only two opposites, it is important to note that the environmental issue for *ecocapitalists* is, according to Lima, an unwanted byproduct of progress, which can be adjustable within the capitalist order, albeit in a selfish and exclusionary way. On the other hand, Marcuse (1967), a representative of *ecosocialism*, as Lima (*op.cit.*) classifies him, stresses the incompatibility between the logic of capitalism and ecology, stating that the Earth cannot be saved within the capitalist system.

According to Lima (*op.cit.*), the construction of the concept of sustainable development began in the 1970s due to the increase in environmental accidents and problems. In 1972, the UN promoted the International Conference on the Human Environment, an initial official landmark in highlighting concerns about environmental issues. In the same year, the Meadows Report (Meadows et al. 1972), after noting the environmental degradation of the planet made pessimistic forecasts, warning about a collapse within a century.

The concept of eco-development was introduced in 1973. Among its basic principles, we can highlight: "...the development of a social system that guarantees employment, social security and respect for other cultures and further education programmes" (Brüseke 1996, quoted in Lima 1997). In 1974, the Declaration of Cocoyoc drew attention to the responsibility of rich countries towards the population explosion, poverty and degradation resulting from the high levels of consumption, waste and pollution.

In 1987, the report *Our Common Future* (or the Brundtland Report), prepared by the World Commission for Environment and Development, contains the following definition of sustainable development: "...the one that meets the needs of the present without compromising the ability of future generations to meet theirs" (quoted in Lima 1997: 14). Lima reminds us that the report raised questions such as: how to achieve economic efficiency, ecological prudence and social justice in a world with extremely unequal, unjust and degraded reality? Are developed countries and the elites of underdeveloped nations willing to change and sacrifice?

The concept of sustainability presented in the Brundtland Report allows, according to Lima, different interpretations. For Herculano (1992) (quoted in Lima, 1997: 17), there is a semantic contradiction, since "sustainability is a term of ecology which means a tendency to stability, dynamic balance and interdependence among ecosystems, whereas development is related to increases in the means of production, accumulation and expansion of productive forces". Another criticism of the concept of sustainability, quoted by Lima, is made by Stahel (1995), who, using the law of entropy as an argument, concludes that the sustainability discourse in the context of a market economy, is an illusion,

> Comparing the significance of biospheric and economic times, noting that the time biospheric is circular, guided by the principle of stability, continuous recycling and by the low levels of entropy, whereas the economic time, introduced by capitalism, is marked by a steady expansion, market competition, by constant innovations and by instability. According to the author, the acceleration of time, a feature of the capitalist logic, disrupts the circular time and the biospheric stability, accelerating entropic degradation processes. A higher productivity and competitiveness represent the generating of high entropy, increasing waste and pollution. In his view, the environmental crisis would be on this time lag between the acceleration of economic time and its inability to adjust to the biospheric time (Lima, 199: 18).

Benfica (undated) writes about terms that have a similar conceptual basis, such as "human development" (Programa Das Nações Unidas Para O Desenvolvimento 2009) which has the advantage of putting human beings at the centre of development, but this has also received criticism. "The concept of human development, whose central axes are 'fairness' and 'participation' is still evolving, and is opposed to a neoliberal conception of development" (p.4). The term, explains Benfica, as used by the United Nations, acts as an indicator of the quality of life, having health, longevity, psychological maturity, clean environment, community spirit and education as indicators of a "sustainable society", which meets current needs without compromising the ability and opportunities of future generations. This proposal is strongly criticised by Altvater (1995: 282) in the text of Benfica, as an "empty formula". However, Altvater agrees that development "should be economically efficient, ecologically affordable, politically democratic and socially fair" (p.5), but he does not see how this can be achieved under the capitalist model, which, as we have seen above, is incompatible with sustainability. Another criticism of the model of sustainable human development, according to Benfica, is that environmentalism treats environmental and social issues separately, as he asserts,

> The success of the ecological struggle will depend greatly on the ability of ecologists to convince the poorest people that it is not only about cleaning rivers, cleaning the air, reforesting devastated areas in order to live on a better planet in a distant future, but also giving a simultaneous solution to environmental and social problems. The problems ecology deals with do not affect the environment alone. They affect the most complex being of nature, the human being (Benfica undated: 4).

Finally, we see in the words of Gadotti (2000: 34) quoted by Benfica (undated) a summary of the current scenario of (un)sustainability, highlighting its inequalities,

> The scenario is given: globalisation caused by the advancing technology revolution, characterised by the internationalisation of production and expansion of financial flows, regionalisation characterised by the formation of economic blocs; fragmentation that divides globalising and globalised ones, centre and periphery, those who die from starvation and those who die from excessive consumption of food, regional rivalries, political, ethnic and religious confrontations, terrorism.

Having the myth of sustainable human development which brings to the few the excesses of pleasure and consumption and to the majority deprivation of all kinds, we are left to reflect on the pedagogical possibilities of preparing a counter-discourse and to try through education, albeit informal, to reverse the neo-liberal logic.

The Pedagogical Discourse

Acknowledging that sustainability is also, and primarily, an educational issue, the question that arises is how to introduce to education the sustainability principle as agreed in 1987. Gadotti (2000) drafted a proposal entitled *ecopedagogy* that starts from such questions as: To what extent is there sense in what we do? To what extent do our actions contribute to the quality of life for people and for their happiness? Gadotti tells us that *sustainable education* is not only concerned with a healthy relationship with the environment, but with the "sense of things from everyday life" (Gadotti 2000: 19). Gadotti's pedagogical proposal involves a reorientation of the curriculum for significant proposals in a broad sense, that is, the one which includes the health of the planet. Gadotti (2000: 26) reminds us that traditional pedagogies are anthropocentric while ecopedagogy departs from planetary awareness (genera, species, kingdoms, formal, informal and non-formal). Indeed, Lampert (2007: 35) reminds us,

> Post-modernity calls for a curriculum that is related to issues of class, race, gender, caste, ideology, individuals, hermeneutics, ecology, theology, cognition and all the "isms" of the "post" era. The postmodern curriculum cannot be defined in terms of subjects and contents. It is a triggering process, of dialogue, of research that seeks transformation. It is a process of exploring the unknown...

The question, however, as Lampert (2007: 17) explains, is: "How to perform a multicultural dialogue when some cultures have been reduced to silence, and their ways of seeing and knowing the world have become unpronounceable? How to make the silence speak without it necessarily speaking the hegemonic language that it wants to speak?"

Two important points arise: social exclusion and digital exclusion. Pennycook, an applied linguist, reminds us that the confluence of language and education refers to "two aspects mainly of life politics, given that societies are unequally structured and dominated by hegemonic ideologies and cultures, which limit the possibilities of thinking about the world and hence on the possibilities of changing this world" (Pennycook 1998: 24). For Pennycook, there is a need for a critical pedagogy in the transformative and emancipatory sense that considers schools as cultural arenas enabling us to "emphasise how subjectivities are constructed within and from the school and how the voice of the student and popular culture are delegitimised forms of culture and knowledge that students bring to school" (Pennycook 1998: 45).

From what has been stated, we find that social exclusion brings and will bring incalculable consequences, including those to the environment. Lampert (2007: 24–25) surveys the damage from the promises of modernity and an exaggerated concern for progress that has been caused to the planet, leaving "mother Earth" in a situation of shortage, arguing,

> ...exclusion and poverty have spread across all continents, reaching mostly the people with less education. The promises of equality, fraternity, justice and human rights, built along the past centuries, are far from being achieved, and for many they are nothing more than utopia (Lampert 2007: 25).

Lampert reminds us that in the 1970s exclusion referred to prisoners, the mentally ill, disabled and elderly. Today its meaning has spread to other layers of society. It is a product of social construction. It has several causes and can affect the citizen in economic, political, social, educational, digital and cultural terms. Lampert adds that violence and peace are social constructs, tied to political and economic life and to social organisation and having a close relationship with education and with pedagogical practices, as violence and peace are taught and learned.

Regarding the presence of technology in the school curriculum, Lampert notes that the schools need to reconcile youth culture with their primary objectives which are the transmission of the cultural heritage and integral formation of the individual under the threat of forming uncritical citizens and rampant consumers, leading to environmental problems. Lampert advocates the use of the internet in education within an educational proposal based on interactivity, customisation, and on autonomous development of learning to think. The schools must be concerned with the humanisation of students and with environmental sustainability and they should take advantage of technological contributions to work critically on the content and the means used. Finally, Lampert (2007: 45) proposes the formation of a citizen of the world who does not renounce his or her cultural roots and who,

> ...measures pleasure with wisdom, learns to work with the losses and uncertainties...A being who knows how to make the other better and happier. A man (*sic*) able to learn how to find time for the family, leisure, the body, pleasure, consumption, rest, love, others, reading, creation, meditation, prayer and solitude...

The *Vozes Que Ecoam* Project

The *Vozes Que Ecoam* project was undertaken during 2008. It was the result of a partnership between the University of Sorocaba, a community higher education institution with a strong tradition in extension activities, the Educational and Vocational Association *Pérola*, a NGO of Sorocaba with activities focused on the vocational training of young people from poor communities, and with the financial support of a local bank and Sorocaba City Hall, through CMDCA (Adolescent and Child City Council).

The *Vozes Que Ecoam* project was initially proposed to the NGO *Pérola* which later resubmitted it to the other partners through public bidding in accordance with the bank and the public agency edict. We chose the NGO *Pérola* as a partner because it is in charge of activities related to 'digital inclusion', which is part of a project with local authorities bringing access to computers connected to the peripheral communities of Sorocaba. For that, 20 buildings had been built, called *Sabe Tudo*, all in the grounds of public schools in poor neighbourhoods. Each building had 20 computers, instructors and offered courses on computer operation and basic principles of navigation and web search, all coordinated by Pérola NGO.

The idea for the project came from reports in local newspapers that boasted that its provision for learning some features of Microsoft Word and some tips on browsing and searching were "digital inclusion". This conception of common sense, that access to computers and the computer network means digital inclusion and brings with it benefits to users, is a Manichean vision and domination practice that results in new forms of control and habits of coexistence with hegemonic uses of language, ways of thinking and consuming. As Jiménez (2008: 21) asserts:

> When culture is mistaken as entertainment, when the public conversation comes down to the last football game, the latest soap opera, to the scandals of artists' lifestyle, recent serial murders, when human life becomes an audience of witnesses of the spectacle of others' lives, what is at stake is civilisation and the need to adjust the order of values to a new society ....

Although this quotation refers to television, Jiménez also criticises the new media, as, for the philosopher, it is also the new pedagogy on which globalisation has been built. He also draws attention to the fact that globalisation, besides not informing, ends up transforming a people's cultural values simply into marketing.

In order to offer an alternate proposal for 'digital inclusion', which would bring benefits to the communities involved, the *Vozes Que Ecoam* project was created. The main objective was to develop pedagogical proposals involving the use of multiple languages with a view to digital literacy. This focused on aspects related to the communities in which participants lived and the recovery of their subjectivity and identity. Also, as part of the project, training for teachers of municipal schools was offered as the same movement could also be articulated from the inside out. The expectation was that digital literacy events would happen in out-of-school time, at 11 Sabe Tudo kiosks available in 2008 (the others were still under construction) and that teachers would develop the same ideas of the project in the school curricula because this would facilitate the dissemination of these new practices of writing in the digital environment and, over time, make these practices official at municipal schools.

The project was developed over 12 months with 560 young people aged between 14 and 24 years, students of municipal schools who had signed up for the project at Sabe Tudo offices. In order to carry out the workshops, we trained 30 instructors with two trainees for educational support. The project was coordinated by two lecturers from the University of Sorocaba, with the logistical support of NGO *Pérola* staff.

The modules were developed in the form of workshops that included discussions and practical activities focused on the social functions of language (especially the verbal and visual), their possibilities for artistic expression and how to undertake a critical reading of the media. We sought to include in the programme *Vozes* the suggestions of Parâmetros Curriculares Nacionais da Língua Portuguesa (Portuguese Language National Curricular Parameters), "...the language, as a construct of the verbal interaction of speakers, cannot be understood without taking into account their link to the actual situation of production. It is within the functioning of language that this operation can be understood. Producing language, one learns language". For the practical activities, including field activity, we used cameras, camcorders and audio and video editing programmes. Participants had lessons on how to operate the hardware and software besides learning about audio and video languages.

The activities offered to students sought to encourage multimodal reading and writing, that is, combining words of written texts with static or moving images and audio sources while developing the skills to use the tools of information and communications technology. The projects developed in the workshops provided the students with the identification of important aspects of local culture, education, health, environment and violence in their communities; field research and trips in their neighbourhoods; socialisation; and dissemination of the final products. It was essential in the *Vozes Que Ecoam* project for participants to become agents of their own history by recording images, texts and videos of the ethos of their communities and local cultures, then making them available on the web through blogs created and maintained by the students themselves.

Among the activities, we highlight the blogs created by students and the project's official blog (http://perolavozesqueecoam.blogspot.com/) that besides linking the students' blogs focuses productions chosen by the students of the participating communities as the most significant. In their blogs, students could express themselves through texts with external links to other texts, images, videos, etc. Over time, it became practical to post the comments of participants from other communities on the work of each of them. Such communication between bloggers was encouraged by activities called "tem não tem" and "história do meu bairro," performed with the aid of Wiki pages and the collaborative production of texts. At first, via posting on blogs, participants compared the needs of their neighbourhoods in terms of safety, education, health, etc. Then each community that had a Sabe Tudo unit was depicted with texts and photographs taken by the students themselves. With that, besides knowing better their origins, their stories were known by colleagues from other neighbourhoods. In addition, the blogs were accessed by people from other cities and states (some even from other countries!) and so the social voices could reach large audiences and expand the social network of these peripheral groups. This occurred in such a way that a student, realising that his neighbourhood did not appear at Wikipedia (http://pt.wikipedia.org/wiki/Jardim_ipiranga), decided to include the data surveyed in the encyclopedia, which was followed by the inclusion of the other participating communities, revealing the cultural identity of peripheral communities and their dwellers.

Examples of interaction among the communities may be seen on the project blog. These are the extracts produced in the Wiki environment:

> Today we are going to play a little. Everyone will write about themselves!!
> Which are your dreams and objectives in life.
>
> Let's go!!

My name is Haphaelle, I am 17 and my dream is to finish College and apply for the Barro Branco Police Academy in São Paulo!!!

I cannot forget about home, of course!!!!!
hahahahahhahaha

I am Fernanda, I am 15 and I study at Flavio [school]. I study at Sabe Tudo, I have a dog, I live with my father and mother, I join the youth jaspion group and I am at the 1st grade of High School.

One of my dreams is to travel to Italy, to the United States and to Japan. Another dream is to join the C.S.I team.

****__my name is Julio, I live in Laranjeiras:

My neighbourhood lacks a Sabe Tudo Unit. I attend the Paineiras Sabe Tudo, that is why I think that there should be a Sabe Tudo in every neighbourhood. The mayor made some changes to the neighborhood but I think they were done only because of the elections as he can be candidate again.****
*jardim ferreira does not have: very little police activity as there are lots of drug dealing points that has been harmful to the children as they do not grow in a healthy environment; There should be surveillance, a lot of under aged drinking alcohol, family schools, on week-ends the schools used to be open for people of the community and take courses;
Jardim Ferreira does not have: a good security for people, Wi-Fi technology for its people to access Internet and optimised police surveillance.

As can be seen in these extracts, the participants are from different neighbourhoods: Laranjeiras, Jardim Ferreira (Paineiras neighbourhood is also mentioned), the writing is strongly marked by orality, there are deviations from the standard norm, elementary spelling errors and interference of Internet colloquial writing; they aspire to goals influenced by hegemonic cultures, notably the U.S., as part of CSI (a criminal investigation television series aired now in Brazil, also in an open channel) and target a training elite, how to join the Academia de Barro Branco. On the other hand, they also show a critical view of problems with alcohol, drugs and violence and lack of Wi-Fi connection. Inserted in mass culture, they desire the cultural and technological products even when they, in most cases, lack the essentials in terms of housing, employment and nutrition.

In the examples of interaction, below, we can see the comments of participants on the results of the work undertaken by students from other communities with regard to the virtual book on the history of each neighbourhood. Interestingly, despite abundant misspellings, the extracts show their pride and finally, an encouraging comment of a person from São Paulo (as evidenced by the IP shown on the blog).

I found this "virtual book" you made very interesting for the people to know the work people do and also our work at Sabe Tudo (Marcelo Augusto) and for getting to know our neighbourhood better.A.de S. M. Jr.

4 September 2008 12:31
I liked a lot to work with Wiki, even more the final result because when we started this topic we did not have an idea of how our work would be, but I am sure that not only me but all students liked this work. Any one who sees it will like it!!!!!!

10 September 2008 10:51
It is very nice how you said about our neighbourhood for people stop with prejudice.

11 September 2008 12:48
O.P. said…
I found it very interesting. It is worth seeing our services we have done to the world. It is very rewarding for us to know about a very good story behind this audiovisual course.
D. C. K. said…
Congratulations for the work! You are building a space where voices and ideas echo. I am astonished with what I have seen and read.

Regards
D. C. K.

Another activity of the *Vozes que Ecoam* project that deserves comment and may have been the most important of all was the production of videos whose scripts, arguments, footage and editing were undertaken by the students in groups in each Sabe Tudo. There are three reasons why this activity is important: first, the complexity of handling the video camera and the software used for editing audio and video; the second reason is that the students themselves chose the subject and approach, i.e., they practised authorship, became producers for the web and were not just consumers of information, and they discussed issues important for them. The third reason is that an audio-visual production must be orchestrated regarding the verbal, gestural, visual and oral language as well as writing and the telling of stories with a video camera. Taking into account the results of the audio-visual production, that contains outdoor and indoor shots, taken within the buildings of Sabe Tudo and from several camera tricks through special effects and also on the variation of angles and planes, we noticed that the students began to understand and use the syntax of audio-visual language and use it to express their feelings and ideas, not only for consumption by websites.

Due to the size of the videos – ranging from 5 to 20 min in length – and quantity, 21 in all, they were not available on the blog but recorded on DVD and shown in open session at the University of Sorocaba. They drew attention to the topics: in *A Surpresa do Caipira* the issue of teenage pregnancy is shown; in *História do Morro* and *Domina Vícios*, for example, the issue is drug dealing; *Little Yellow Riding Hood* and *Rapunzel* retell familiar stories but with plots adapted to the students' reality, the scenarios are the vacant lots of the neighbourhoods and the castle of Rapunzel is the building of the community Sabe Tudo. Class and colour prejudice is treated in the videos *Contraste Social* and *Preconceito*. Criticism of the communication media, TV in particular, was not missed in works like *TV Chinfurifula* and the News *Jornal Pode Tudo* made in a satirical and mocking way. Violence appears in *A Emboscada*. Other interesting topics were *Mais Amor Menos Descaso*, *O Verde no Mundo*, *Alunos Idiotas*, besides less critical work, such as *Amantes do Skate* and *Moda Emo*.

The *Vozes que Ecoam* project was completed in December 2008 with the issuing of certificates to the participants in the presence of their families in a club located in the central square of Sorocaba. This was an important point as it works as a metaphor of voices echoing from the periphery to the centre.

Conclusion

With this work we tried to show that the concept of sustainable development and its maintenance has come down to us through a discourse that emanates from the hegemonic powers. It is not a coincidence that these powers are those less interested in curbing consumption since it would imply restrictions on what they had to offer for sale with a consequent reduction in industrial production, leading to unemployment and a decrease in revenues. It seems to us that the boasting of the media regarding sustainability in terms of environmental preservation (preservation of the ozone layer, garbage collection, and global warming, etc.) is nothing more than a way of drawing attention to very important aspects, of course, but whose 'solution' would not impact on the hegemonic powers and the great corporations in the same way that restrictions on consumption would. Consumption is the central problem.

On the other hand, we emphasise that it is not enough either to include ecology as a subject in the school or in programmes for planting tree seeds on celebration dates or even to strengthen in the curriculum such topics as energy generation, energy consumption and energy savings. These topics are important but do not reach the core of the sustainability problem: a life lived for consumption.

In order to offer a counter-discourse we sought in Gadotti and Lampert and Pennycook, among others, a critical pedagogy for resistance, for the ecology, for the Earth. This pedagogy, as we have seen, provides for the formation of an ethical, balanced, happy and peaceful citizen. It is not a spiritual pedagogy in the religious sense of the term, as it may seem at first glance. It is a pedagogy that considers the other in any space and at any time, present or future, for we are all part of the same ecosystem.

The *Vozes que Ecoam* project was not proposed explicitly to discuss sustainability. We followed what Lampert (2007) advises, that is, that our curriculum does not originate in metanarratives but comes from experience and local history. We have presented a proposal contrary to the hegemonic model of usage of information and communications technology which does not teach or provide new uses. By contrast, it tends to maintain the hegemonic practice that has a devouring impulse, as Demo (2008) warns, for whom new technologies have an economic significance that exceeds by far its social significance. "Our main function continues to be consumption" (Demo 2008: 5). Citing Meszários (2002), Demo makes a cruel observation: we seem more disposable than the technologies, even though they also echo the liberal market, because everything becomes a commodity and so do we. So, without metanarratives and against the dominant practices of the discourse, language and technology are used to speak of the subjects and their communities, making the voices echo from the periphery to the centre.

The *Vozes que Ecoam* project developed voices as an out-of-school project with a small, but not innocuous inclusion in the classrooms of those teachers who also attended the course. But it is clear that one year is such a short time and the time outside school is not sufficient. We know that our youth is still shopping at the "cultural supermarket" (Mathews 2002) where there is available information and identities, but perhaps many of them are doing it in a more critical and aware way. As Kleiman and Vieira (2006: 119) assert,

> …it is key, in any society that the individuals learn in schools to defend themselves, against influences consciously and critically, adverse or not, which act through new discursive practices imposed by the use of the computer. If this individual is not prepared to select critical information, he *(sic)* will be an easy target for all sorts of manipulation.

Participation and the intervention of schools are extremely necessary for these new practices related to the use of writing in the digital environment. The access and the use of technologies have been taking place through communities of practice in informal settings, internet cafes and at work. The appropriation of language, messages and ways of thinking occur in these environments and are influenced by these environments, which make more urgent the need for schools and educators to be updated about the usage of new media. As Machado (2006) asserts, "In the future, when the cameras will replace the pens, when the computers will edit movies instead of text, those probably will be the way to "write" and give shape to our thinking". On the role of schools in these new times, Kleiman and Vieira (2006: 30) suggest that,

> …if technology changes go unnoticed by the educational institution, the school is in danger of becoming invisible and missing the low prestige that has been left. Society needs, urgently, a school that does not place itself at the margin of the world, being neglected by it or being seen as benignly complacent.

The project blogs are currently on the web; the city First Lady has been given a CD of the work of the project. We went from the periphery to the central club to present the participation certificates; the project has been the subject in the most watched TV news in town. In short, the voices echoed, but the political gains have not yet been felt. But we know that there have been individual and social gains. The testimonies of the students show us that some participants found jobs because of the skills developed in the course. We believe that virtually everyone involved learned different uses of the computer, including looking at themselves and their communities.

Going from the periphery to the centre also provokes resistance. We offered the project to the NGO *Pérola* in 2009, with some modifications and requirements, but could not continue with it. We wanted the computers of *Sabe Tudo* to have connections that would support videos, not only text. We wanted to have cameras and video recorders and for the *Sabe Tudo* Units to have a more frequent maintenance (because they would not be available due to flooding and falling wires). We also wanted them to be more accessible for disabled people, since *Sabe Tudo* buildings were not designed for people with physical disabilities and there are no computer programs accessible to the visually and hearing impaired.

We have proposed to continue working with the alumni through interaction activities among other connected Brazilian peripheral communities, such as the CUFA (Central Única das Favelas www.cufa.org.br) with a new group organised for municipal school teachers. Another proposal was to post the blog in English as well. We wanted to have conversations in Portuguese and in English with communities from other countries. At the end, we did not have the project accepted in the terms that we had proposed and we ended up working almost on our own with a new group of teachers from the city schools. As we prepare reformulations and opportunities for our project, we believe that the voices still echo.

References

Altvater, E. 1995. *O Preço da Riqueza: Pilhagem Ambiental e a Nova (des)Ordem Mundial*. São Paulo: UNESP.

Armitage, J. 1999. Resisting the neoliberal discourse of technology. In *The politics of cyberculture in the age of the virtual class*, eds. A. Kroker and M. Kroker, Available in httpw://www.ctheory.net/articles.asp?id=111 (accessed February 17, 2010).

Benfica, G. (undated). Sustentabilidade e Educação. *Revista da Faculdade Estadual de Educação da Bahia*. Salvador: UNEB. 10(16): 13–25.

Bruseke, F J. 1996. Desestruturação e desenvolvimento. In *Incertezas de Sustentabilidade na Globalização*, ed. E Viola and L C Ferreira. Campinas: Unicamp.

Buarque, C. 1990. *A Desordem do Progresso: O Fim da Era dos Economistas e a Construção do Futuro*. Rio de Janeiro: Paes e Terra.

Demo, 2008. *Habilidades do Século XXI*. Boletim Técnico Senac: *a* Revista Educação Profissional. Rio de Janeiro, 34 no.2, (May/August): 4–15.

Gadotti, M. 2000. Pedagogia da terra e cultura de sustentabilidad. *Revista Lusófona de Educação* 2005(6): 15–29.

Hall, S. 2005. *A identidade Cultural na Pós-Modernidade*, 10th ed. Rio de Janeiro: DP&A Editora.

Herculano, S C. 1992. Do desenvolvimento (in) suportável à sociedade feliz. In *Ecologia, Ciência e Política*, ed. M Goldenberg. Rio de Janeiro: Revan.

Jiménez, J. L. F. (2008). Tecnocultura y desempleo: las tecnologías del entretenimiento base pedagógica de la globalización Boletim. Técnico Senac: *a* Revista Educação Profissional. Rio de Janeiro, 34 no.2 (May/August): 16–25.

Kleiman, A B, and J A Vieira. 2006. O impacto identitário das novas tecnologias da informação e comunicação (Internet). In *Práticas Identitárias: Língua e Discurso*, ed. I Magalhães, M Grigoletto, and M J Coracini. São Carlos: Claraluz.

Lampert, E. 2007. Pós-Modernidade e educação. *Linhas* 8(2): 4–32.

Lima, GF. da C. 1997. O debate da sustentabilidade na sociedade insustentável. *Política e Trabalho* no.13 (September): 201–222.

Machado, J. 2006. *Os computadores na facilitação da aprendizagem: estudo tomando o conceito de função*, Dissertação de Doutoramento. Braga: Universidade do Minho.

Marcuse, H. 1967. *Ideologia da Sociedade Industrial*. Rio de Janeiro: Zahar Editores.

Mathews, G. 2002. *Cultura Global e Identidade Individual*. Bauru: EDUSC.

Meadows, D H, D L Meadows, J Randers, and W W Behrens III. 1972. *The limits to growth: A report for the club of Rome's project on the predicament of mankind*. New York: Universe Books.

Meszários, I. 2002. *Para Além do Capital*. São Paulo: Boitempo Editorial.

Pennycook, A. 1998. A Linguística Aplicada nos anos 90: em defesa de uma abordagem crítica. In *Linguística Aplicada e Transdisciplinaridade: Questões e Perspectivas*, ed. S Inês and C C Marilda. Campinas: Mercado das Letras.

Programa Das Nações Unidas Para O Desenvolvimento. 2009. http://www.pnud.org.br/home/ (accessed February 17, 2010).

Sachs, I. 1986. *Caminhos Para o Desenvolvimento Sustentável*. Rio de Janeiro: Garamound.

Stahel, A W. 1995. Capitalismo e entropia: os aspectos ideológicos de uma contradição e uma busca de alternativas sustentáveis. In *Desenvolvimento e Natureza: Estudos Para uma Sociedade Sustentável*, ed. C Cavalcanti. São Paulo: Cortez.

Chapter 14
At School You Learn About the City: Urbanisation and Sustainable Development in Education

Paulo Celso da Silva

Introduction

> Smooth is the night
> At night is when I go out
> To know the city
> And get lost
> Our city is very big
> And so small
> So distant from the horizon of the country
>
> (Engenheiros do Hawaii 1986)

The song 'Longe demais das Capitais', from the gaucho group Engenheiros do Hawaii (1986) starts one more geography class in a Paulista state school. From memory, all of the students sing it. After the final chords are sounded, the students applaud and ask for another song, they want one more. However, it is now time to move to an analysis, though it is better not to throw away or lose this moment of euphoria for the students.

What Is the City?

This question refers to a complex of possible responses that pervade the history of mankind. We can begin to find an answer in the nineteenth century when the sciences as we know them today were systematised. Thus, Humboldt (1966), in 1811, studied and analysed Mexico City and its urban fabric based on the Aztecs.

P.C. da Silva (✉)
University of Sorocaba (UNISO/SP), São Paulo, Brazil
e-mail: paulo.silva@prof.uniso.br

M.L. de Amorim Soares and L. Petarnella (eds.), *Schooling for Sustainable Development in South America*, Schooling for Sustainable Development 2,
DOI 10.1007/978-94-007-1754-1_14,

His analysis was thorough because it was not limited simply to the economic and population data. He also tried to analyse the pre-Hispanic civilisations that he admired. Engels (1985), studying Manchester in England, claimed in 1845 that the city is about agglomeration. Catalan Ildefons Cerdà (1867), creator of the word urbanisation, said that formation is nothing, the satisfaction of needs is everything. In Austria, Camillo Sitte (1992) in 1889 proposed the city as a work of art, with small squares to prevent agoraphobia. Ebenezer Howard (1996), in turn, created the concept of the Garden City with its 12 ha of garden in the city centre. On the other hand, Tony Garnier (1988), in 1901 reflects on the industrial city located near the raw materials and modes of transport with 25,000 inhabitants accommodated in houses on plots of 15×15 m, each divided in half with a building and a green area. Georg Simmel (1903) examined the relationship between the metropolis and mentality, arguing that the small city is related to feelings and sensitivity and the big city is related to reason. Max Weber (1950) examined the medieval city as a crowded habitat, a large town where size is not the distinguishing feature because you need to see the contents.

On the other hand, Raoul Blanchard (1911) analysed those geographical factors, including the physical and human elements, which explain why a city is located in this spot and not in another, considering that localisation is more important than the spot, exemplifying this with Argel and its important position in the Mediterranean. Well known for his buildings, the architect Le Corbusier (1998) said that mules do not walk straight but men do, and cities are built for the mules, asserting: "My proposal is brutal, because urbanism is brutal".

Kevin Lynch (1960) argued that the city exists more than sensibility can see. The city is the object of the perceptions of its inhabitants, it is the result of the process of observations. More concretely, Jean Remy (1994) affirms the city is a unit of production and distribution and a consumer of knowledge. The Catalan Carles Carreras (1993) analysed Barcelona, affirming that the city is a palimpsest, a text constantly rewritten. Another Catalan, Francesc Solé (2001), affirmed that the "city is a sum of agents, inter-related and connected with the outside within its territory. The element of coherence and coordination of ability to manage the knowledge with profits, it is now local and global".

The contemporary city can be seen as a mixture of all of these references, with fragments of past times which mark the view, the roughness that stayed inside new tissues which form and establish themselves, creating a space in the city. Viewed this way, the city offers itself as a book to be read, to be interpreted and interpenetrated throughout the school environment, including those environments that are considered rural. More than theorists of cities, students create mental maps and a cartography that is distinct from official maps and guides. Students map the city based on their experiences, preferences and readings. The street where they live is the centre of the district and the district is the centre of the city. In this way, there always exist new centres since students or groups of students take turns in the spotlights that the school world place on their students according to the occasion. Hence, perception is always growing; it means each one receiving the perception of the other and moving the centres in the district. It is clear that the school is a privileged locality in which to live, to feel, and to understand, to learn the city, no matter if the order of the phenomena is random. The important thing, though, is that the city is experienced as a happening.

There are signs indicating that the city constantly demonstrates new and old situations: traffic jams, pollution, mobility, uneven growth, irregular consumption, an unequal city. At school, it is possible, and necessary, to learn the signs as one learns about other cities. Entering European castles, watching the green of the Forest People, understanding a Spanish pueblo with only 50 people, hearing the white of the Yakutsk in Siberia with its −50°C with only a few weeks of summer, surf the "big arteries and veins" that are the streets of large cities and/or many Brazilian rivers or in the Mississippi Delta, enter into the development of Lausanne in Switzerland, pursuing the heat with the Bedouins...The possibilities of the other cities are endless, as well as our own city.

> Smooth is the night
> At night is when I go out
> To know the city
> And get lost
> Our city is very big
> And so small
> So distant from the horizon of the country
>
> (Engenheiros do Hawaii 1986)

Experience has shown and contemporary practice demonstrates that individuality in the city can be a problem for all, seen, for example, in the rise in the number of individual vehicles. The same practice has also shown that collective solutions help to reduce the problems in the city as, for example, in the search for sustainable development. Together, city residents produce millions of tons of trash and residues everyday that need a destination. The residues can be recycled, thus being able to be returned to social use in the shape of other products and the trash must have its own destinations and it remains in the society which produced it. This simple conception has immediate implications for the life of many people and for decisions that may or may not favour collectivity and sustainability.

Accepting that sustainable development is defined as having the ability to meet the needs of the current generation without compromising the capacity for attending to the needs of future generations, as a development that does not exhaust resources for the future, we already have the conditions to identify the importance of schools as transformation agents for the immediate collective reality. Thus, if the tone of many cities, the habitat of the great majority of people in the world, is individualism as a mark of urbanity, urban schools are the locus to find other ways to live and socialise people. We have one more reason to find viable solutions to sociability and conscious occupation. Observing the criteria of the European Energy Award (EEA) to award cities with the seal of rational energy use, we see that a range of activities including territorial development, municipal buildings, power supply, mobility, internal organisation and communication parameters can apply to any city, whether in Europe or elsewhere. These activities, taught in schools, certainly have a big impact in the long term on local society because all of the items are likely to be taught and discussed within the school community. Indeed, the very space of the school can be used for citizen discussions and for the processes of public consultation which aim to promote public participation in the process of decision making in governmental actions.

I have always wanted to live in old world
In the old way of living
The 3rd sex, the 3rd war, the 3rd world
Are so difficult to understand

Smooth is city
To whom like the city
To whom have needs to hide

(Engenheiros do Hawaii 1986)

Latin American Data

The so-called 'third world countries' were named this way by the demographer Alfred Sauvy during the Cold War and adopted in 1955 at the Bandung Conference held in Indonesia by Asian and African countries who had fought for independence from the colonising European countries. They aimed at economic and cultural cooperation between themselves, confronting both the U.S.A. and the USSR. It was the first conference to affirm that imperialism and racism were crimes. It is important to remember the meaning of the term Third World as proposed by Sauvy. It was based on the French Revolution experience with the Third State, where the third world should make the revolution as the French bourgeoisie had done. The world was understood as being divided into blocks: the first was the developed capitalist world; the second was the socialist world; and the third world was undeveloped and poor.

This conception, however ideological and purposeful, lasted until the collapse of the Soviet Union and today it is preempted by such terms as developed countries, developing and/or emerging countries. However, there is considerable conceptual difficulty in determining exactly the stage of a country's development. The World Bank (2010), for example, uses a strict numerical criterion where those countries with a GDP per capita (gross national income (GNI) per capita) below US$11.905 and above US$900 are considered emerging, according to 2008 data. Table 14.1 shows low-income groups, US$975 or less; lower middle income, US$976–US$3,855, upper middle income, US$3,856–US$11,905, and high income, US$11,906 or more.

The World Bank ranks only 29 Latin American and Caribbean countries as having economies higher than the average shown in Table 14.1 These Latin American Countries are: Argentina, Ecuador, Panama, Belize, El Salvador, Paraguay, Bolivia, Grenada, Peru, Brazil, Guatemala, St. Kitts and Nevis, Chile Guyana, St. Lucia, Colombia, Haiti, St. Vincent and the Grenadines, Costa Rica, Honduras, Suriname, Cuba, Jamaica, Uruguay, Dominica, Mexico, Venezuela RB, Dominican Republic and Nicaragua.

It is interesting to study the economic indicators, shown in Table 14.1 for the first decade of the twenty-first century. One of them is population which indicates a growth to half a million inhabitants. Referring to the data shown in Table 14.2, from

Table 14.1 Latin American economic indicators 2002–2008 (Source: http://www.latin-focus.com/latinfocus/countries/latam/latpopulation.htm)

Latin America							
Economic indicators, 2002–2008							
Real sector	2002	2003	2004	2005	2006	2007	2008
Population (millions)	500.4	507.5	514.0	519.8	526.9	532.3	541.2
GDP per capita (US$)	3,426	3,637	4,155	4,993	5,811	6,790	8,112
GDP (US$ billions)	1,714	1,846	2,136	2,595	3,062	3,614	4,390
GDP (annual growth in %)	0.6	2.0	5.9	4.6	5.5	5.7	4.3
Unemployment (%)	8.3	8	7.6	6.7	6.9	6.3	6

Table 14.2 Population changes in Latin America 2000–2010

	2000	2010
Population	516 millions	576 millions
Urban population	389 millions	471 millions
Rate of urbanisation	75.30%	79.00%

the Estado das Cidades da América Latina e do Caribe, the last published numbers by Agência para Habitação das Nações Unidas (ONU–Habitat) in March 2010, it can be seen that there was a total of 576 million inhabitants, of whom 471 million people were living in cities, that is, 79% of Latin Americans, making this the most urbanised region in the world. It is interesting to notice the changes that have occurred in the first decade of the current century (see Table 14.2).

The publication *State of the World's Cities 2008/2009: Harmonious Cities* (United Nations 2008: 263–264) (http://www.unhabitat.org/pmss/listItemDetails.aspx?publicationID=2562) indicates possible population growth patterns for Latin America and the Caribbean, from which we quote some Brazilian cities in Table 14.3.

From a reading of the reported data, it becomes evident that city schools have a number of problems that they can use in teaching about sustainable development. In Rio de Janeiro, the World Social Forum, that took place between March 22 and 26, 2010, addressed such topics as: finding ways to reduce disparities in access to the city; reducing inequality and poverty; access to decent housing; adequate sanitation; and the provision of urban services for all. Also included were discussions on youth and the city, since young people are the most numerous demographic group in the developing world. The World Social Forum held in Rio was the first to participate in the Global Urban Campaign led by the United Nations which aims to include sustainable urbanisation in the policy agendas at all geographic scales, from the local to the global. This opens up another debate that has been unresolved by globalisation.

Our city is so small
And so naive
We are so far
From the capitals

(Engenheiros do Hawaii 1986)

Table 14.3 Estimated population and projections in thousands of selected Latin American cities

Cities	2000	2010	2020	2025
Belém	1,748	2,335	2,639	2,717
Belo Horizonte	4,659	5,941	6,597	6,748
Brasília	2,746	2,938	4,463	4,578
Campinas	2,264	3,003	3,380	3,474
Cuiabá	686	857	972	1,008
Curitiba	2,494	3,320	3,735	3,836
Florianópolis	734	1,142	1,328	1,374
Manaus	1,392	1,898	2,156	2,223
Natal	910	1,161	1,316	1,362
Porto Alegre	3,505	4,096	4,517	4,633
Recife	3,230	3,831	4,236	4,347
Rio de Janeiro	10,803	12,171	13,179	13,413
Salvador	2,968	3,995	4,113	4,222
São Paulo	17,099	19,582	21,124	21,428

Globalisation and the City

Globalisation has become a recurring theme in school curricula. It is a theme that is difficult to ignore since it is continually referred to in the mass media to designate many processes occurring in daily life. Globalisation is not one process which everybody can say that they have enjoyed since the 1980s that marked the end of, the so-called, bipolar world. Since then, the neo-liberal policies of the great powers of the USA and the USSR have been attractive to some countries and have been 'imposed' on others, mainly those that had previously been called 'underdeveloped' and are now described as 'emerging countries'.

In an effort to define globalisation, the economist Alvaro Antonio Zini Jr. affirmed,

> ...globalisation is the acceleration of contacts, exchanges and international travel. These have always existed and, by themselves do not characterise that process again. But the speed of trade has grown exponentially over the past ten years. Globalisation is the expansion of exchanges between people and companies in different countries. Three sets of factors give fuel to this expansion. The first is the technological revolution brought about by the computer, which increased the ability to process information for individuals and large companies. The cheapening of information transmission and long distance national and international comes up next. Finally, the decrease in the price of international transport and the increase in international travel adds to the list. Meanwhile, globalisation has caused great perplexity, for fear that it would cause unemployment, reduction of state sovereignty and more widespread use of popular culture (Zini Jr. 1996: 2–10).

The fear, to which he refers in the final sentence of this quotation is linked to the exhaustion of the constitutional state and privatisation. In South America we can point to the privatisation of such basic industries as water supply as in Bolivia in 2000 resulting in confrontations between the police and civil society where people demanded the de-privatisation of water supply. In the same year there were confrontations arising from the nationalisation of the gas industry in Brazil. In 2001 there was a crisis in the Argentine economy calling for many economic and financial solutions resulting in what Santos called "perverse globalization" where money and information tyranny compose the "ideological base that legitimise the actions" (2000: 37).

Some theorists and major international media affirm – perhaps with less emphasis today than in the 1990s – that globalisation is like a natural law and is therefore inevitable. Some voices in the same major media disagree. In 1997, the writer Carlos Heitor Cony questioned this 'naturalisation' of the economy, in a short article entitled *Gravity and Globalisation*. Here, he argued, that it was not "exactly man's struggle not to abolish the law of gravity, but disciplining it for his own benefit... Human civilisation is a struggle against the natural forces. Invoke this kind of force to justify globalisation is to return humanity to the age of the cave" (1997: A-2). In the same journal there was in 2003 an editorial titled *End of Globalisation*, affirming that, "The main economic victim of the U.S. offensive in Iraq can be globalisation. For the most pessimistic, even the failure of globalisation has been the cause of this military action" (2003: A-2).

Whatever the media, the information passed to the great majority of the population is the same in all places in the world. The producers of the mass media select what is or should be news, since few agencies distribute the same stories to the media worldwide. Hence there is the sense of a Global Village where more is known about what happens around the world than in the surroundings of our homes, our district or our city. There can be no doubt that information and communication technologies have made a qualitative leap, including equipment with increasingly lower prices, such as camcorders and computers. However, the programming of broadcasters or the contents of the print media have not always followed this qualitative leap in the way they present the world and thus they have been preparing a standardised spectator.

Analysing how globalisation has progressed in the world, and looking specifically at the reality experienced in Latin America, Santos (2000:18–21) identifies the facets of globalisation as,

1. The world as they make us believe: globalisation as fable;
2. The world how it is: globalisation as perversity;
3. The world as it can be: another globalisation.

His analysis concluded with a proposal for a more human globalisation that is more connected to those who are excluded from the 'capital mentality,' where people cannot see a contradiction in having congested traffic while the majority of cars carry only one person. In the same vein, people do not see a contradiction when a person dies because he or she cannot pay for the specialised health treatment that is required. Santos makes an appeal to the conscience in the interest of future generations. We talk too much today of progress and the promises of genetic engineering that would lead to a mutation of biological man, something that is still in the domain of the history of science and technique. Little, however, is said about the conditions that can provide a changing philosophy of man, able to give a new meaning to the existence of each person and the planet (2000: 174).

The analyses of Pankaj Guemawat on the global economy indicate that there is not, in fact, globalisation and that is why the idea of semi-globalisation is more appropriate to the times in which we live. It is argued that we are a long way from reaching the absolute integration that we should seek during the next decades (2008: 27). Guemawat (2007) focuses his efforts on indicating to executives the importance of differences between countries, instead of emphasising standardising them as if boundaries or peculiarities no longer existed. In an interview for the newspaper *La Gaceta*, online edition (Guemawat 2007), it is affirmed,

> Just a week ago I was interviewed by Fareed Zakaria, a well-known journalist from Newsweek, and he asked me: "What is your image of semi globalisation, a shocking word?". The reason why this word is used is due to the fact that globalisation has not really come. The Financial Times has just published a review of my book, which he defined as "a skeptical view of globalisation." What follows this view is that if you do not believe, as I do not believe, in a 100% cross-border integration, you are a skeptic, because that is globalisation. This view of complete integration is considering that it is erroneous or partial, and this is the reason I use the word semi globalisation. Because, even for such an approach advocated by the Financial Times, globalisation means a complete integration of borders, and that is an idea with which I disagree, because I do not think that the borders are fully integrated.

His book is written as an academic study of the internationalisation of companies. When we consider the differences between places, companies would provide this internationalisation but the commitment exists in order to ensure greater profits and sales success of the brands offered in the market. There is no concern for social or moral responsibility in the territories occupied by multi- or trans-national companies. For most companies, a charge diverges and skips completely the logic based on the requirement of internationalisation.

The relationship between globalisation and the city is a dialectical relationship where the global profits and advances of some companies are gained at the cost of the misery and poverty of the many. We can refer as well to the relationships between the global and the local, considering that the city is the locus where the daily life of the majority of Latin Americans occurs. In the introduction to the book published by the World Bank (2002), *Globalisation, Growth and Poverty: Building an Inclusive World Economy*, it is stated that it is unacceptable for one fifth of the global population to live with less than a US dollar per day in a world which produces more and more, a world of abundance that, as already demonstrated by many specialists, produces more food than humankind could consume. Certainly, the majority of these people live in cities.

Such people and their poverty when concentrated in cities are visible to the naked eye. A tour of the major urban centres of Latin America would verify that dignified living conditions – incidentally, a right that is guaranteed – are not receiving a high priority. Returning, to the history of Brazilian urbanisation, we verify that the process of urbanisation received a great impulse with the coming of industrialisation and modernity. However, the cities retained the spatial and social arrangements for the excluded that had persisted from the time before capitalist models were introduced. More and more people have migrated from the countryside to the cities bringing increasing demands for basic urban services such as health, education, energy, water, transport and housing.

In general, several modes of production exist together in the city. In other words, handicrafts and manufacturing coincide in time and space with, for example, the capitalist textile companies producing in the city but the products being purchased by the people in rural communities who were there at the very start of the textile production process. The industrialisation process in Brazil has intensified mainly since the 1930s. However in several cities its genesis must be sought in the latter half of the nineteenth century. Often, the process of industrialisation was directly related to coffee production, building the coffee industry where investments in industry were made by the producers of coffee when the product price was low, thus, they sought to diversify their investments. However, it seems more appropriate to think about industrialisation as a process involving actors and products besides coffee since in some parts of the southern state of São Paulo cotton rather than coffee was one of the drivers of industrialisation (Martins 1990: 106).

It was not only coffee production that supported the growth of cities. The city of São Paulo, for example, had its industrial development process more pronounced in the late nineteenth and early twentieth century, times when the city underwent landscape reforms, improved basic sanitation works, improvement of transport routes and also the formation of working-class districts on the outskirts. Here industrial plants were located including the Santana Factory, Alvares Penteado, Francisco Matarazzo and Crespi, among the best known. A little of the life of the workers in the 1920s and 1930s was recounted by Patricia Galvão, then code-named Mara Lobo under pressure from the Communist Party, in the proletarian romance *Industrial Park* (1933). She denounced the exploitation of the workers and condemned their poor living conditions. New workers' villages were built providing a model of living and architecture that marked many Brazilian industrial cities.

In his work entitled *The City in History*, Lewis Mumford (2008) presents Coketown and his description is relevant for almost all working-class districts and the industrialisation processes that formed them,

> Industrialism, the main creative force of the nineteenth century, produced the most degraded urban environment that the world had ever seen...The basic politics of that new kind of urban aggregation was based on three main pillars: the abolition of the guilds, with the creation of a state of permanent insecurity for the working classes; the implementation of an open labour market and competitive, as well as the fair sale of goods.
>
> Perhaps the most important fact of all in the urban transition was the shift of population that occurred throughout the planet...The main elements of the new urban complex were the factory, the railroad and the beehive. The factory became the nucleus of the new urban organism. All other details of life were subordinated to it...The plant usually demanded the best sites (Mumford 2008: 531–573).

At the same time that the workers lived in the Brazilian industrial suburbs, the capitalist real estate market indicated those areas which would be valued and those not requiring landscape improvements. Also, Mumford referred to the poor aesthetic quality, a result of the economic situation, further impoverishing the environment in which the workers lived. Not only was the outside chaotic and ugly, inside the houses there was "...furniture inspired by the worst examples of bad taste by the middle class: the futility of uselessness" (2008: 555).

Another problem in the situation of the working classes in Brazil is that since the institution of the Law of Lands in 1850, which organised and normalised the urban and rural lands, only lands that were bought and sold were legal before the law. In practice, this apportioned different values to different parts of the city resulting in a considerable proportion of the population seeking areas less valued and less accessible and with fewer urban amenities. This exclusionary model created both a legal city and an illegal city. The first is urbanised and legalised by the government and the second comprises 'invasion' or 'squatter' areas that are illegal and with no basic infrastructure. This model has persisted to the present day.

The right to the city is a fight which gathers many social actors and many institutions. Quoting again from the Global Urban Forum, in the words of Anna Tibaijuka, Deputy General Secretary of the United Nations and Executive Director of ONU-HABITAT,

> When we say the right to the city, we are talking about ensuring that women, men, youth and children have equal access to basic services in the communities where they live. These basic services include access to safe water and adequate sanitation, so people can live with dignity in an environment free of disease. The right to the city also means minimum levels of security, so that people do not live in constant fear of suffering assaults or robberies. The right to the city also includes energy and accessible public transport to facilitate access to employment, education and leisure. The right to the city includes the right to adequate housing and the right of individuals to participate in decisions that affect their means of livelihood. At last, the right to the city should mean equal opportunity for everyone to improve their living conditions and livelihood without putting at risk the rights of future generations to do the same (2010: 3).

Education, the City and Sustainable Development

> Smooth is the night
> At night is when I go out
> To know the city
> And get lost
> Our city is very big
> And so small
> So distant from the horizon of the country.
>
> (Engenheiros do Hawaii 1986)

The geography class at Paulista public school is at end, students have heard the bell to signal the end of the lesson and it is time to move to another class. But the song continues to play and to show the possibilities of living in the city which can be small, close or far from the capital. The three terms that prompted the class – education, the city and sustainable development refer to daily and future habits which require big accomplishments. Each achievement creates a demand for new achievements: achievements of the right to the city and to the present/future quality of life with the hope of advancing socially in the community.

The new generations, with the vast quantity of available information presented in several media and gained through social electronic networks in which they participate, want from education an ally for the development of their personal information filters and groups that help them in producing, circulating and consuming information. Each educated person is a new creator of aware opinions.

Awareness in the city is an awareness which seeks the development of everybody. It is an awareness that asks questions of everybody: Sustainable development is sustainable for whom? What are the limits of sustainable development? Pollution, is it to be obliterated or managed? Does capitalism have a scale of unsustainable development or not? These are questions that cannot be answered but require reflection. However to stimulate this reflection the questions must be posed daily and not only in geography lessons. Hence the importance for students to develop filters with the assistance of teachers. School knowledge is important in countries like Brazil where inequality reaches dangerous levels as the headline from the newspaper *O Estado de S. Paulo* from March 2010 shows, *Goiânia is the most inequal city in Brazil.*

Five Brazilian cities are among the 20 most unequal in the world. A Report (O Estado de S. Paulo 2010, p.C-4) presented at the opening of the Fifth World Urban Forum of the United Nations (UN) in Rio shows that Goiânia (10th), Belo Horizonte (13th), Fortaleza (13th), Brasília (16th) and Curitiba (17th) are those with the greatest differences in income between rich and poor in the country. It also reports that Brazil is the country with the greatest social distance in Latin America. Rio, in the 28th position, and São Paulo, in the 39th, are also cities with high inequality, according to the report. Analysing the whole country, the report show that in Brazil 50.6% of income is concentrated in the hands of the richest 10% and the poorest 10% get only 0.8% of Brazilian wealth. This confirms the fact that Brazil is the country with the highest rates of inequality in Latin America.

Also, reflection is needed on cultural issues and on education and sustainable development. Many people need to live with respect, confidence and communication because globalisation means that people who can (having money, personal energy, family help, for example) move to places with good conditions. For so many years Brazilian cities have been the destinations for many Latin American, African and Asian families seeking a place to develop their potential. Although we have no xenophobic problems, we should not believe that our racial problems are fully resolved.

We believe that cities are privileged places of study and fellowship. The first decade of the century showed that one-way roads, homogeneous – as they said about globalisation – are not the most appropriate to look at the diversity that people in the world offer, diversity that makes each country a surprise to be discovered, such a surprise that international trade agreements must respect it to keep the benefits. These first years of the century present solidarity. More and more people believe that quality education is coming to the poor people to bring the will of socialisation, in other words, those who have school experience fight for those who have not yet had the opportunity.

References

Blanchard, R. 1911. *Grenoble, Etude de Géographie Urbaine*. Paris: Armand Colin.

Carreras, C. 1993. *Geografia Urbana de Barcelona. Espai Mediterrani, Temps Europeu*. Vilasar de Mar: Oikos-tau.

Cerdà, I. 1867. *Teoría General de la Urbanización, y Aplicación de sus Principios y Doctrinas a la Reforma y Ensanche de Barcelona: La Urbanización Considerada Como un Hecho Concreto: Estadística Urbana de Barcelona*. Madrid: Imprenta Española (Madrid).

Cony, C. H. 1997. *Gravidade e Globalização*. Folha de S. Paulo, A-2.

Corbusier, L. 1998. *Por uma Arquitetura*, 5th ed. São Paulo: Editora Perspectiva.

de Martins, J S. 1990. *O Cativeiro da Terra*, 4th ed. São Paulo: Hucitec.

Engels, F. 1985. *A Situação da Classe Trabalhadora na Inglaterra*. Trans. Rosa Camargo Artigas and Reginaldo Forti. São Paulo: Global.

Engenheiros do Hawaii. 1986. *Longe demais das Capitais*. (Compact Disc) Brasil: RCA.

Folha de S. Paulo. 2003. *End of Globalisation*. Editorial, A-2.

Garnier, T. 1988. *Une Cité Industrielle: Étude pour la Construction des Villes*. Paris: Philippe Sers.

Guemawat, P. 2007. *Está Lejos la Globalización, Muy Lejos. Nosotros no la Veremos*. www.gaceta.es/25112007+pankaj_ghemawat_esta_lejos_globalizacion_muy_lejos_nosotros_no_veremos,noticia1img,36,36,490. (accessed May 20, 2008).

Guemawat, P. 2008. *Redefiniendo La Globalización. La Importância de las Diferencias em um Mundo Globalizado*. Barcelona: Deusto.

Howard, E. 1996. *Cidades – Jardins de Amanhã*. São Paulo: Hucitec.

Lobo, M. 1933. *Parque Industrial. Edição Fac-similar*. São Paulo: Alternativa.

Lynch, K. 1960. *The image of the city*. Cambridge: MIT Press.

Mumford, L. 2008. *A Cidade na História, suas Origens, Transformações e Perspectivas*, 5th ed. São Paulo: Martins Fontes.

O Estado de S. Paulo. 2010. *Goiânia é a cidade mais desigual do Brasil*. March 22, 2010. C-4.

Rémy, J, and L Voyé. 1994. Contexto urbanizado e efeitos de estrutura social. In *A cidade: rumo a uma nova definição? Tradução de José Domingues de Almeida*, ed. J Rémy and L Voyé. Porto: Afrontamento.

Santos, M. 2000. *Por uma Outra Globalização. Do Pensamento Único à Consciência Universal*. Rio de Janeiro: Record.

Simmel, G. 1903. As grandes cidades e a vida do espírito. *Mana* [online]. 2005, Vol.11, no.2: 577–591.

Sitte, C. 1992. *A Construção das Cidades Segundo seus Princípios Artísticos*. São Paulo: Editora Ática.

Solè, F. 2001. Revista Barcelona Metròpolis Mediterrànea – Ciutat del Coneixement. *Els Monogràfics No. 1*, Ajuntament de Barcelona.

UN-HABITAT. 2009. *State of the World's Cities 2008/2009. Harmonious Cities*. (First published by Earthscan in the UK and USA in 2008 for and on behalf of the United Nations Human Settlements Programme). www.unhabitat.org (accessed February 9, 2010).

UN-HABITAT. 2010. *Quinta Sesión del Foro Urbano Mundial. El Derecho a la Ciudad: Uniendo el Urbano Dividido. Dialogue*, Trabaje en Red, Aprenda y Exponga en la principal conferencia sobre ciudades, en Rio de Janeiro, días 22 y 26 de marzo de 2010. www.unhabitat.org/wuf (accessed February 12, 2010).

von Humboldt, A. 1966. *Ensayo Político Sobre el Reino de la Nueva España, (Estudio preliminar de Juan A. Ortega y Medina)*. México: Porrúa.

Weber, M. 1950. *La Citta. Trad. O.Padov.*. Milano: Ed Valentino Bompiani.

World Bank. 2002. *Globalisation, growth, and poverty. Building an inclusive world economy*. Washington, DC/Oxford: A co-publication of the World Bank/Oxford University Press.

World Bank. 2010. *Country Classification*. http://web.worldbank.org/WBSITE/EXTERNAL/DATASTATISTICS/0, contentMDK:20420458~menuPK:64133156~pagePK:64133150~piPK:64133175~theSitePK:239419,00.html (accessed February 9, 2010).

Zini, Jr, Á. A. 1996. *Globalização*. Folha de S. Paulo, 2–10.

Index

A

Active citizenship, 24, 35, 39
Afonso, C.M., 9
Agriculture, 9, 35, 53, 62, 98, 123, 131
Altvater, E., 205, 208
Alves, U.S., 53
Amazon, 3, 16, 20, 118, 159–172, 175, 176, 190
Andrade, J.B.T., 7
Anti-nuclear groups, 21
Aragón, L.E., 168
Argentina, 15, 50, 62, 65, 71–84, 222
Armitage, J., 206
Autonomy, 3–17, 95, 128, 130, 146, 151, 155, 156, 165, 199–201

B

Bacon, F., 20
Ball, S.J., 120
Barbrook, R., 154
Baudelot, 106
Bauman, Z., 20, 194, 202
Becker, B.K., 164
Begatini, R.M., 150
Benejam, P., 71
Benfica, G., 55, 205, 208
Bertelli, M.D., 150
Bilingual education, 101
Biocidade (biocity), 12
Biodiversity, 3, 8, 14, 34, 79, 112, 114, 161, 164, 169, 171–173, 175, 183, 193
Biotechnology, 25–26
Blanchard, R., 220
Blog, 212–214, 216
Boff, L., 55–57, 152
Bolivia, 14, 15, 53–68, 96, 222, 224
Bordenave, J.E.D., 149
Boschetti, 14
Boschetti, V. R., 19
Bourdieu, 106
Bowe, R., 120
Brazil, 5–16, 26, 62–66, 68, 105–109, 112, 113, 117–120, 123–141, 143–156, 159, 160, 164–166, 169–171, 182, 183, 185–188, 195, 200, 206, 213, 222, 225, 227–229
Brazilian Agenda 21, 9
Brazilian Institute for Geography and Statistics (IBGE), 135, 170, 171, 176
Buarque, C., 206

C

Camargo, A. L de B., 23
Capitalism, 19, 23, 105, 107, 112, 126, 144, 162, 169, 196, 207, 208, 229
Carbon emissions, 131
Carreras, C., 220
Carril, C., 162
Cartographic materials, 38, 48
Carvalho, I., 4
Carvalho, I. C de M., 21
Carvalho L.H., 146
Carvalho, V.S.A., 109, 111, 120
Castro, G., 96
Catalão, V.L., 27
Cavalcante, M.E.S.R., 165
Cavalcanti, L., 68
CBC. *See* Common basic contents
Cerdà, I., 220
Cerrado (tropical savanna), 159

M.L. de Amorim Soares and L. Petarnella (eds.), *Schooling for Sustainable Development in South America*, Schooling for Sustainable Development 2, DOI 10.1007/978-94-007-1754-1, © Springer Science+Business Media B.V. 2011

Chile, 14, 33–51, 222
Citizen formation, 33, 34, 37, 38, 48, 51
Citizenship, 3–17, 24, 28, 29, 35, 39, 49, 64, 89, 95, 98, 108, 112, 120, 126, 131, 132, 139, 146, 151, 152, 156, 172, 173, 182, 184, 187, 189, 196
Civilisation crisis, 193
Claure, K., 58
Climate, 8, 9, 12, 34, 43, 53, 62, 114, 131, 165
Coca, 15, 61–64, 67
Coelho, M.C.N., 163
Collective identity, 54
Common basic contents (CBC), 75, 79
Community, 5, 16, 50, 56, 59–62, 64, 84, 90, 92, 94, 100, 108, 110, 113, 115, 116, 120, 127, 130, 131, 133, 139, 149–154, 156, 160, 164, 174, 182–184, 186, 187, 190, 192, 197, 199, 201, 208, 210, 212–214, 221, 228
Conservation, 6, 10, 26, 79, 93, 110, 112, 114, 149, 161, 169–172, 175, 206
Consumers, 21, 62, 63, 80, 145, 187, 194, 210, 214
Consumption, 6, 9, 10, 16, 21, 23, 63, 80, 112, 114, 132, 133, 144–148, 154, 156, 187, 193–195, 202, 206–208, 210, 214, 215, 221
Cony, C.H., 225
Corbusier, L., 220
Costa, C.A.F., 198, 199
Costa Neto, C., 62
Crespo, S., 9
Critical thinking skills, 35
Cross-curricular, 15, 26–28, 74, 75, 77, 82, 105–120, 131, 133–135, 200
Cultural diversity, 12, 35, 40, 57, 61, 111, 119, 132, 159, 161, 164, 169, 171, 172
Curriculum design, 72, 83, 90
Curriculum development, 24, 46, 48
Curriculum implementation, 38

D

de Amorim Soares, M.L., 3, 105
Decade of Education for Sustainable Development 2005–2014, 4, 13, 34, 71, 72, 120
de D'Amico, R.L., 87
Deforestation, 11, 20, 26, 63, 71, 131, 151, 161
Degradation, 8, 10, 17, 73, 75, 114, 162, 169, 182, 195, 200, 202, 206–208
Deleuze, G., 198
Demographic dynamics, 40
Desertification, 35, 71
Design Possível Project, 181–192
Development indicators, 40
Dias, G., 110
Digital inclusion, 17, 206, 210, 211
Digital literacy, 205–216
Disciplinary approach, 77
Diversity, 3, 7–10, 12, 14, 22, 26, 27, 34, 35, 40, 57, 61, 64, 77, 79, 92, 97, 111, 112, 114, 117–119, 132, 159, 161, 164, 169, 171–175, 183, 193, 195, 199, 229
Domingues, J.M., 63
Drought, 35

E

Early childhood education, 16, 101, 143–156
Earth Charter, 8, 114, 146
Earth's tectonic plates, 44
Eckersley, R., 113
Eco–92, 8, 9, 12, 13, 53, 131, 169, 195
Eco-development, 7, 22, 28, 169, 187, 207
Ecological ethics, 198, 199
Ecology, 7, 22, 25, 28, 29, 182, 183, 186, 187, 201, 207–209, 215
Economic growth, 7, 9, 21, 76, 113, 164, 167, 168
Eco-pedagogy, 55, 64, 209
Ecosocialism, 207
Educational pedagogic resource, 181–192
Educational reforms, 15, 33, 107–110
Educational system, 15, 35, 48, 61, 73, 78, 82, 83, 88, 93, 95, 96, 108, 111, 137
Education transformation, 87–101
EEA. *See* European Energy Award
E-learning, 30, 46, 48
Elementary education, 34, 61, 130
Elias, N., 54
Employment, 7, 64, 66, 88, 207, 213, 228
ENAD. *See* National Examination for Performance Evaluation
ENDS. *See* National Strategy for Sustainable Development
Energy resources, 25
Engels, F., 220
Environmental balance, 5, 39, 171
Environmental crisis, 4, 22, 193, 208
Environmental education, 3–6, 12–14, 22, 26–28, 33, 47, 48, 72–75, 77–81, 83, 84, 90, 93, 94, 100, 101, 108, 110–114, 118–120, 123, 173, 175, 176, 182, 183, 188, 195, 196
Environmental ethics, 198, 199
Environmental journalism, 26

Environmental management, 25, 48, 77, 78, 84
Environmental policy, 5, 40, 45, 73, 77–79, 83, 84, 113
Environmental protection, 5, 9, 22, 72, 81, 112, 182
Environmental quality, 5, 23, 75, 183
Environmental rights, 93
Environment and Sustainable Development Olympics, 81
Equilibrium, 20, 23, 24, 98, 117, 124, 132
Establet, 106
Ethical motivation, 35
Ethical relationships, 198–201
Ethnocentrism, 68
European Energy Award (EEA), 221
Exploitation, 19–21, 62, 71, 72, 77, 112, 141, 145, 164–166, 170, 171, 187, 201, 206, 227
Extinction, 20, 53, 123, 141, 153, 193, 206

F
Federal law, 5, 74, 75, 83
Feminism, 113
FGV. *See* Getulio Vargas Foundation
FNEP. *See* National Federation for Private School
Food and Agriculture Organisation (FAO), 53, 123, 131
Formal education, 16, 24, 46, 48, 54, 59, 71, 77–84, 88, 109, 111, 156, 173
Fossil fuel, 9, 71, 114
Foster, J.B., 19
Foucault, M., 196
Freire, P., 199
Frers, C., 72
Fresh water, 123

G
Gadotti, M., 27, 28, 55, 56, 64, 205, 208, 209, 215
Gallo, S., 197
Garbage, 11, 13, 24, 26, 27, 54, 115–117, 126, 152–154, 193, 206, 215
Garnier, T., 220
Gas, 9, 22, 61–64, 67, 114, 123, 131, 132, 225
Gatarri, F., 195
GDP. *See* Gross domestic product
Geographical education, 33–38, 47–51, 72, 74
Geographic space, 14, 37–41, 43–46, 82
Geosystem, 41, 44
Getulio Vargas Foundation (FGV), 137
Glaciers, 131
Global development, 74, 163
Globalisation, 14, 17, 20, 23, 34, 35, 37, 40, 45, 51, 61, 65, 74, 82, 96, 146, 168, 183, 202, 208, 211, 223–229
Global warming, 14, 20, 37, 56, 114, 131, 215
Golub, S., 10
Gonçalves, M.A.R., 21
Google Earth, 50
Greenhouse effect, 9, 53
Green schools, 94
Gross domestic product (GDP), 62, 99, 169, 222, 223
Guemawat, P., 226
Guimarães, A.S., 63
Guimaraes, L.B., 21

H
Hage, J.A.A., 62
Hall, S., 206, 210
Hawaii, E.D., 219
Hegemonic social model, 202
Herculano, S.C., 207
Herremans, I.M., 35
Higher education, 33, 35, 60, 91, 96, 98, 99, 101, 129, 130, 139, 140, 159–176, 210
Howard, E., 220
Human development, 34, 35, 55, 56, 167, 190, 196, 208
Human development index (HDI), 167, 190
Human-factor-consumption, 145
Humanistic, 15, 92, 95–97
Humankind, 19, 23, 29, 110, 111, 226

I
IBGE. *See* Brazilian Institute for Geography and Statistics
Immigrants, 3, 65, 66, 68, 126
Inclusion, 5, 15–17, 29, 75, 88, 93, 97, 101, 108, 111, 113, 138, 175, 182, 184, 186, 188, 192, 206, 207, 210–212, 215
Individual identity, 54
Individualism, 56, 67, 111, 112, 200, 221
Industrialisation, 14, 19, 21, 65, 148, 154, 226, 227
Industrial revolution, 41, 144, 145, 147, 154–156, 193
Inequality, 11, 23, 36, 89, 99, 106, 133, 207, 223, 229
Information and Communication Technologies (ICT), 47, 49, 51
Infrastructure, 9, 10, 76, 89, 90, 96, 106, 117, 125, 165, 169, 170, 228

Interdisciplinary approach, 4, 110, 202
International Union for Conservation of Nature and Natural Resources (IUCN), 6
Internet, 47, 51, 206, 210, 213, 216

J
Jacobi, P., 28
Japiassu, H., 29
Jiménez, J.L.F., 211
Jimenez, S., 22, 24

K
Kleiman, A.B, 216
Krajcberg, F., 118

L
Lampert, E., 209, 210, 215
Land, 16, 19, 35, 43, 59, 62, 64, 124–27, 146, 149, 151–156, 162–164, 168, 202, 227
Latin America, 65–68, 81, 87, 88, 96, 164, 184, 223, 225, 226, 229
Learning, 4, 10, 14, 24, 29, 30, 33, 36– 48, 54, 57, 58, 61, 64, 65, 75, 83, 89, 91, 95, 97, 98, 105, 111, 113, 119, 129, 131, 133, 143, 147–149, 152, 155, 156, 184, 191, 197, 198, 200, 205, 206, 210–212
Legal Amazon, 161, 163–166, 171, 172, 176
Leo Maar, W., 151
Lima, A.T., 118
Lima, G.F., 206, 207
Livestock, 62
Lobo, M., 227
Loreto, M., 87, 89
Lynch, K., 220

M
Machado, J., 216
Manzini, E., 181
Manzochi, L.H., 196
Marcuse, H., 207
Marginalisation, 21
Martins, J.R., 173
Marx, K., 19, 144
Mazochi, L.M., 146
Mazzotti, T.B., 195
Meadows, D.H., 207
Melo, C.K., 173
Mendoza, O., 87
Meszários, I., 215
Mining, 25, 65, 75, 80, 168, 170, 222
Mission Robinson I, 88, 98
Mission Robinson II, 88, 98
Monteiro, M.A., 163
Mota, Á., 63
Multiculturalism, 15, 68, 97
Multidisciplinary approach, 77
Multiethnicism, 15, 97
Mumford, L., 227
My Neighbourhood, My Land, My Treasure Project, 16, 146, 149, 151–156

N
National Curriculum Parameters (PCN_s), 15, 105–120, 152
National Examination for Performance Evaluation (ENAD), 138
National Federation for Private Schools (FNEP), 137
National Strategy for Sustainable Development (ENDS), 75, 76
National System for Environmental Certification of Educational Centre (SNCA), 47
Natural hazards, 36
Natural resources, 6, 9, 11, 19–21, 23, 25, 29, 36, 64, 67, 68, 71, 72, 75–77, 79, 112, 113, 144, 145, 147, 152, 154, 159, 162, 164, 165, 168, 170, 172, 181, 187, 201, 206
Neoliberal policy, 67
Neto, M.J., 164, 165, 167
Ngo Florescer, 184
Noal, R.E., 19
Non-formal education, 78, 80, 81, 84
Non-governmental organisation (NGO), 7, 8, 12, 13, 54, 96, 113, 114, 131, 182, 184
Non-renewable resources, 21, 148

P
Paes, J.P., 201
Pages, J., 74
Palacios, F.A., 33
pantanal (wetlands), 159
Papaneck, V., 183
Paradigms, 21, 24, 155, 172, 173
Passeron, 106
Pato, C., 27
Paulo, S., 227, 229
Peace, 6, 12, 29, 56, 61, 67, 91, 92, 98, 113, 116, 205, 210, 215
Pelicioni, M.C.F., 195

Pennycook, A., 209, 215
Per capita income, 167, 169
Person-consumer, 145
Person-statistician, 145
Petarnella, L., 3, 105, 143
PETISCEN. *See* Strategic Plan for Information Technology and Communication in the National Education Sector
Philippi Jr., A., 182
Piaget, J., 199, 200
Pitano, S de C., 19
Planet, 13, 14, 17, 19, 20, 22, 24, 27, 30, 37–39, 41, 44, 46, 53, 54, 56, 62, 67, 68, 112, 117, 119, 123, 131, 139–141, 146, 150, 153, 163, 169, 188, 193, 198, 199, 201, 202, 207–210, 225, 227
Planetary civilisation, 64
Planting, 58, 62, 63, 115, 215
Plurilingualism, 15, 97
Polimeni, C.M., 71
Political education, 3, 128, 182
Pollution, 11, 20, 21, 25, 36, 53, 56, 62, 71, 73, 78, 113, 118, 131, 201, 207, 208, 221, 229
Pons, I.E.R., 181
Population, 4, 5, 10, 11, 15, 21, 23, 36, 40, 43, 45, 57, 58, 65, 71, 88, 89, 93, 95, 123, 125, 126, 129, 135, 136, 138, 139, 141, 145, 151, 152, 160, 161, 165, 167–171, 176, 190, 193, 207, 220, 222– 228
Poverty, 5–7, 9, 12, 22–24, 36, 56, 74, 76, 89, 106, 133, 140, 169, 206, 207, 210, 223, 226
Preservation, 10, 20, 24, 25, 27, 64, 77, 112, 113, 117, 133, 137, 145, 150, 152, 154, 170, 175, 215
Project Escola do Amanhã (Schools for Tomorrow Project), 114–116

Q
Qualitative methodologies, 48
Quantitative methodologies, 48

R
Racism, 68, 195, 222
Reflection, 4, 9, 17, 19, 27, 35, 53, 94, 108, 118, 123, 131, 146, 147, 150, 151, 153, 155, 181, 188, 196, 198–200, 202, 229
Regional development, 163, 165–172, 174, 175
Reigota, M.A., 6, 21, 25
Remote sensing, 25, 74
Ribeiro, D., 124, 130
Rural development, 35

S
Sachs, I., 7, 22, 187, 206
Sá, L.M ., 27
Santos, J.S., 165
Santos, M., 20, 225
Saviani, D., 105
School culture, 123–141
School dropout, 93, 137
Schramm, F.R., 162
Scotto, G., 21
Secondary education, 33, 34, 36, 46, 49, 59, 61
Self-determination, 5, 91, 92, 198
Semi-arid, 159
Serres, M., 193
Silva, S.A., 63, 64
Simmel, G., 220
Simoncito (Little Simon Project), 88, 98
Sitte, C., 220
Slum, 5, 116, 126, 127, 133, 184
SNCA. *See* National System for Environmental Certification of Educational Centre
Soares, M.L., 143, 144, 150
Social equity, 75, 189, 192
Social exclusion, 93, 162, 209
Social *habitus*, 54
Social integration, 99, 130
Social justice, 3–17, 23, 61, 88, 91, 134, 187, 207
Social security, 7, 88, 93, 207
Social values, 4, 113
Socio-economic changes, 55
Soil, 22, 25, 53, 63, 65, 71, 118, 131, 170
Solidarity, 4, 5, 7, 9, 34, 58, 61, 89, 91, 92, 95, 100, 152, 156, 188, 195, 199, 205, 229
Sorrentino, M., 182
Spatial distribution, 39, 41, 44
Spatial dynamics, 38–40, 45
Stahel, A.W., 207
Strategic Plan for Information ? Technology and Communication in the National Education Sector (PETISCEN), 99
Sugar cane, 26, 118, 125

T

Teacher training, 12, 33, 34, 48, 60, 89, 90, 97, 130
Teaching-learning process, 45, 48, 75
Technology, 6, 7, 9, 17, 24–26, 49, 60, 61, 65, 71, 88, 97, 99, 101, 111, 119, 144, 152, 162, 166, 173, 174, 182, 188–190, 192, 193, 208, 210, 212, 213, 215, 216
Terceiro, E., 22, 24
Textbooks, 46, 47, 49
The Federal University of Tocantins (UFT), 159–176
Third world, 113, 222
Toxic chemicals, 63
Trafficking, 116, 150
Transdisciplinary, 57, 74, 75, 135
Trends in International Mathematics and Science Study (TIMSS), 96
Tropical forests, 123

U

UFT. *See* The Federal University of Tocantins
Unemployment, 11, 65, 137, 215, 223, 224
United Nations (UN), 4, 7, 9, 13, 22, 34, 35, 47, 55, 56, 71, 93, 111, 113, 120, 123, 131, 162, 164, 208, 223, 228, 229
United Nations Educational, Scientific, and Cultural Organisation (UNESCO), 12, 22, 34, 47, 60, 71, 72, 96, 110, 184
United Nations Institute for Training and Research (UNITAR), 13
University, 16, 29, 35, 46–50, 59–61, 71, 88, 93, 94, 97–99, 101, 115, 118, 159–176, 181, 183, 185, 186, 189, 205, 210, 211, 214
Urbanisation, 10, 14, 21, 37, 71, 219–229

V

Vásquez, V.H., 57
Venezuela, 15, 50, 62, 65, 87–101, 222
Vieira, L., 162
Viola, E.J., 113
Violence, 82, 106, 115–117, 193, 198, 200, 210, 212–214
von Humboldt, A., 219
Vozes que Ecoam (Echoing Voice) project, 17, 205–217
Vygotsky, L.S., 196

W

Water supply, 10, 152, 224
Weber, M., 220
World Commission for Environment and Development (WCED), 72
World Wide Web (WWW), 51

Z

Zini, Jr, Á.A., 224
Zitzke, V.A., 168